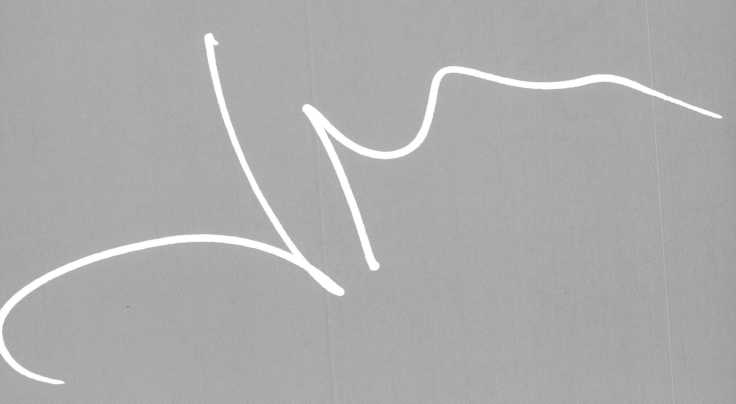

鹹 蝦 燦

味道日誌

給我的家人，DLDL
我的學生們，我的粉絲們。

To my family, DLDL
To my cooking students, and my fans.

煮

Cooking

遊

Travel

序　Prologue

　　《鹹蝦燦味道日誌》記錄了在過去 300 天曾經發生過的事情。我接觸過的人物、到過的地方，這些軼事帶給我不同的思緒，而思緒總會令我回想起一些「味道」的感覺。

　　我是一個感性重於理性的人，從小到大對於身邊發生的事情、遇到的人物都會帶給我一些聯想空間，細味背後的意義。我曾經有擔心自己的「文筆」會否嚇人一跳？內心一直在糾纏著「日誌」是否正確的方向。開始把有點兒雜亂無章的日誌組織過來，再看一看，感覺好像樣啊！

　　我期望這本日誌和食譜帶給讀者正能量的衝擊，喚起大家的一些思緒。這種衝擊對於活在大時代的我們來說是非常重要。到了今天，之前的那種矛盾和糾纏不清的感覺已經拋開了。為甚麼？可能這本書可以寫的也差不多寫完了，所以既來之，則安之。

　　這本書也教曉了我「慢下來」的深層智慧。以往的 2 年是「進取期」，踏進了烹飪界，成為了料理導師，學習拍片、剪片，帶同學到外國烹遊之旅，不停的衝、衝、衝。《鹹蝦燦之味》是一本半自傳式的食譜書，「味道」伴隨我成長，中西生活文化互相交替下造就了我對澳葡菜結下永不分開的「食緣」，這個「食緣」加深了我對家人的認識和憶懷之念，也連結了我跟無數朋友的「友情」。《鹹蝦燦味道日誌》是《鹹蝦燦之味》的延續，這本書走進了我生命裡其中 300 多天的空間，有喜的，也有悲的；有充滿活力的，也有無聊的。不論是晴天或陰天，入廚的過程原來是我最嚮往的事情。

　　入廚的時候，一切都慢下來、靜下來。很多事情既然是不太清楚，就要學懂慢下來，想一想，然後再慢慢的走，一步一步的走，這種感覺是值得回味的。能夠在我人生的第四十個年頭再次為自己的著作寫下一篇「序」實在是太奇妙了！是的，我確實是一個非常幸運的人。

2019 年 8 月 16 日

I've always believed in a strong connection between food and people. Food connects people, bridging our gaps, and creating wonderful memories. That is the rationale behind this book, where I invite everyone to join my journey, recalling what happened in the past 300 days—the people I met, the things I did, and the places I visited. All these little episodes open the door into my cooking world, a world full of taste, love, laughter, tears — or just stupidity. Yes, I am a lucky one!

August 16, 2019

團隊的話　From the Team

這是我第二次跟 John 合作，我的工作主要是負責英文校對和編輯，希望將這本書「活出來」。每次校對完成 John 寫下的食譜，我都想立刻入廚 ；而每次看見他外遊的種種見聞，我都有股立刻買機票的衝動。John，非常榮幸和愉快跟你一起創作這本書。

This is the second of John's cookbook I have worked on. It has been such a joy to work closely with him, once again, on the English portions and recipes, helping him finesse and fine-tune his English-language with "voice." Every recipe I edited made me want to immediately head to the kitchen and start cooking, and every journal entry made me want to book the next flight out. John, it's been a pleasure and an honor to work with you again from halfway around the world.

Deri Reed
Food Editor · New York City

嘴饞的我跟隨著一個對煮食充滿熱誠的廚師，再加上一班合作無間的出版團隊，烹調出這本有人情味的食譜，希望大家都和我們一樣享受到「鹹蝦燦」的甜酸苦辣，隨心隨意的人生點滴。

The best thing that could ever happen to a foodie like me is to be a part of the team publishing John's second cookbook, inspired from his Portugal background and journal entries. Hope you all enjoy it as much as we do.

Jenny Fung
Managing Editor · 香港人出版

這是第二次和 John 合作了，經過去年差不多四個月的合作，這次更加有默契和信任，製作也更嚴謹和仔細，希望能在設計上延續 John 的烹調風格及生活品味，讓讀者更了解他如何把生活細節及經驗都放進菜式裡。很期待吧！

I worked with John for almost four months on his first book last year, and with this second book we have even better synergy and better trust. And this time, we have set a higher standard and expectations. I hope my ideas have helped John to translate his colorful culinary world and tasteful lifestyle into a great book. I just can't wait to see it!

Matthew
美術總監 Art Director

第一次接觸 John 應該是在 2018 年「家電 · 家品 · 博覽」的廚藝示範環節，他絕對是在做一個 show，吸引了很多現場觀眾，還記得當時有一位觀眾在跟她的朋友說：John 是「超班馬」！之後在料理班見過他，那一場料理班是我聽過最多歡笑聲、一場物有所值的料理班。今次正式跟 John 合作，對他更有深入認識，他對每一件事，拍攝的鏡頭、觀眾的反應完全想得非常仔細。不要看他平常嘻嘻哈哈，他絕對非常清楚在做甚麼。非常感謝 John 邀請我參與這本書的製作。

The first time I met John was during a cooking show at "Home Delights Expo 2018". He was on the stage and his cooking performance was so good that it attracted a lot of people's attention. And then I attended one of his cooking classes, and it was the best I have had. Everyone was laughing and enjoying his class. I am very honored to be part of the team and work with John intensively as his videographer. He is very particular with the details and he knows what he wants.

Nick M

視頻製作及策劃 Videographer & Producer

　　很高興能夠成爲是次《鹹蝦燦味道日誌》製作團隊成員之一，從第一次見面，整個籌備及製作過程中都感受到 John 對其作品的認真，希望各位讀者都能夠從書中的相片及文字中感受到 John 設計每道食譜背後的故事及理念！

I am so thrilled to be part of the team. I could feel John's passion and dedication about the book from the very first day we met. I hope everyone loves the journal entries and thoughts behind each recipe that are conveyed through the photos and words.

周偉雄

攝影師 Photographer

目錄 Content

第二部 再煮 Part Two More Cooking

第三部 繼續煮 Part Three
Cooking Is Never Ending

後記
Afterword

煮

Cooking

新書出版、簽書會，
出外旅遊，散心。
旅遊的意義，
沒有爸爸的首個聖誕節，
牽掛、放心。
開心、不開心，
煮，
是治癒的。

I'm not a chef. But I'm passionate about food ——
the tradition of it, cooking it and sharing it.
~Zac Posen~

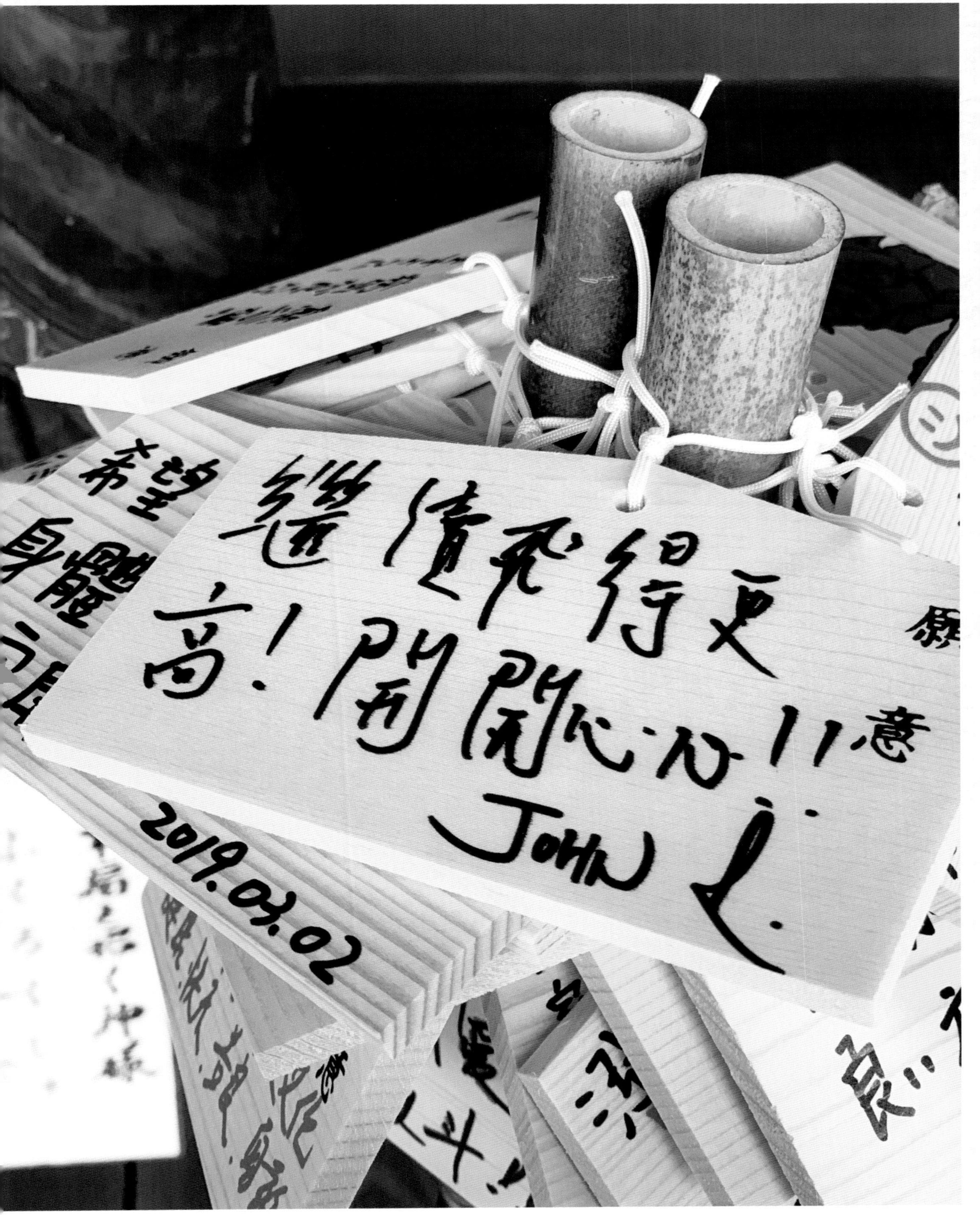
繼續飛得更高！開開心心！！
John L.
2019.03.02
希望
身體

隨身攜帶　A Pen for Signing

我喜歡背著一個包包，放了不可缺少的 power bank 「尿袋」，上堂用的圍裙、記事簿、鉛筆、銀包、零零碎碎的紙張、家人送給我刻了「Chef John」的 Atelier Cologne 和一本放在包包卻久久都還未有空看的書。這兩星期，我的包包多了一支用來簽名的筆。是啊！我現在已經成爲了一名 author，有自己的著作了！

上星期到戲院看戲，竟然有 fans 跟我打招呼，還說知道了爸爸的事替我難過。回到教室，同學們拿著我的書，叫我給他們簽名。我走到鄰近 Towngas Cooking Centre 的書店，見到自己的書竟然上了排行榜！我靜靜的站在一角，看看有沒有人拿起《鹹蝦燦之味》。

開心！當然是開心啦！

你哋就給我多一、兩星期的「沾沾自喜」吧！這支隨身攜帶用來簽名的筆是我付出了很大努力而換來的。But 一向跟我要好的戰友爲甚麼一直 play stranger ？

人與人的認識是一種緣份，不論是朋友，又或是敵人。在上班的環境中，我們除了有一班經常約埋一起飯局的好同事，也一定會有些總令我們神憎鬼厭的同事。要遠離他們？想開一點，我們的認識也是緣份吧。

寫到這裡，想起「昔日戰友」，不如做一道葡萄香草雞吧！

It's been almost two weeks since the launch of my first book, and butterflies are still fluttering in my stomach. I love going to a bookstore, standing in a corner, and peeping to see if anyone picks up my book, looks through it, and heads to the cashier. When I travel to cooking classes, I usually carry my backpack with apron, notes, and notebook—and now I even have a special pen for signing my book in case the occasion arises.

Gosh… Why am I sounding so "vain" ? Hmmmmm… maybe "vain" is a bit too over the top… How about simply letting myself indulge in this state of mind for a little longer? Do you know I have worked so hard to become who I am? And I do deserve this.

And I know this is just the beginning. I need to work extra hard, and even harder to become better and better. I know there are people who love me, and who support me; but there are also people who are jealous of me: All things are relative and in the end I don't care much, as I am not here to please everyone.

I want to enjoy this moment in my life, then get ready for more challenges ahead! The pen I bought for signing books has a very deep and special meaning to me. I worked so hard for it.

Alright, my journal is getting a bit lengthy and it's time for some action. I need a celebration. And I want to cook. I want a dish that requires some chopping and baking… I look in my fridge and see that I've got a chicken and some fresh herbs. Okay… I know what I want to do.

葡式提子百里香煮雞
Portuguese-Style Chicken with Grapes and Thyme

我在其他食譜書見過類似的葡式提子煮雞，有些會用葡萄乾代替提子。但我較為喜歡用新鮮紅提。紅提的甜味與軟腍腍的雞肉很配，湯汁簡單味美。

I have seen similar Portuguese chicken and grape recipes in other cookbooks, with some using raisins instead of grapes. But please try to use red grapes as I am sure you will simply fall for it. The sauce itself is simply just yummy!

材料（2人份量）

橄欖油	2茶匙
雞腿	2隻
胡椒粉	½茶匙
海鹽	½茶匙
無核紅提	250克
乾蔥	2粒，一開四
乾百里香	½湯匙
雞湯	100毫升
新鮮百里香葉	1湯匙，切碎

Ingredient (Makes 2 servings)

Olive oil	2 teaspoons
Bone-in chicken legs	2
Ground pepper	½ teaspoon
Sea salt	½ teaspoon
Red seedless grapes	250g
Shallots, quartered	2
Dried thyme	½ tablespoon
Chicken stock	100ml
Fresh thyme leaves, chopped	1 tablespoon

做法 How to cook

預熱焗爐至180°C。預備一個可入焗爐的煎鑊，加1茶匙橄欖油以中大火燒熱。雞腿灑上胡椒粉和鹽，將雞腿放入鍋中煎8分鐘至金黃，期間反轉一次，離火備用。

在一個碗中，放入1茶匙橄欖油、提子、乾蔥和乾百里香拌勻，然後放入煎鑊中，再倒入雞湯。將煎鑊放入焗爐，焗20至30分鐘。只要用叉插入雞肉內試熟，如流出清澈的湯汁而非血水，即代表雞肉已熟。

將提子及乾蔥平均勻地舀在雞上，以新鮮百里香葉裝飾，即可享用。

Preheat the oven to 180°C. In an ovenproof skillet, heat 1 teaspoon of the olive oil over medium-high heat. Sprinkle the chicken with pepper and salt. Add the chicken to the skillet and cook for 8 minutes, turning once, until browned. Remove the skillet from the heat.

Ready a bowl, mix the remaining 1 teaspoon olive oil with the grapes, shallots, and dried thyme. Spoon the grape mixture around the chicken in the skillet and add the chicken stock. Transfer the skillet to the oven and bake for 20 to 30 minutes, until the chicken is cooked through. To check, poke a fork into the chicken. If it is easy to poke through, and the juices that come out are clear, the meat is cooked and ready to serve.

To serve, spoon the grape mixture and shallots evenly over the chicken and garnish with fresh thyme.

泰葡一族 · Kudichin

自從 2 年前於曼谷 Blue Elephant Cooking School 認識了其創辦人 Chef Nooror，她向我介紹了泰國菜與葡國菜的關係，我便不停在網上做了很多的資料搜集，這也促成了我舉辦 2 次的曼谷之旅。

葡萄牙女士 Dona Maria del Pifia 於百多年前將雞蛋、蔗糖、椰奶和麵粉等食材傳入泰國宮廷料理內，而泰國人也跟葡萄牙人通婚，他們的後人也開始聚居於曼谷一個名爲 Kudichin 的小社區。

今天在 Okura Prestige Bangkok 酒店吃完早餐，我和同學們一行人去到這個鮮爲人知的社區打卡參觀，還可以嚐到泰葡料理。

Bean Sakurthong 是一間小型家庭式經營的小館子，也是區內可以嚐到泰葡料理的地方。Bean Sakurthong 的食譜就像我家中的食譜，是一代傳一代，充滿了歷史和懷舊的味道。我們吃盡不同的菜式，而其 Kanon Jeen 菜式就跟澳葡家庭料理中的葡國雞有點相似，我也突然想起用這菜式再配合嫲嫲葡國雞（《鹹蝦燦之味》第 73 頁）做一個新派葡國雞，味道應該會不俗啊！最有趣的地方就是店主的丈夫長得不像泰國人，是一個 typical 的「鹹蝦燦」look。我不禁想起「when 鹹蝦燦 meets another 鹹蝦燦」。

Kudichin 座落於曼谷市一個小小的 corner，沒有太多人對它有認識，其獨特的飲食文化是不是一段被人淡忘了的歷史？不過我卻自私一點，寧願這裡繼續不要給人留意，這樣便可以保持一份寧靜和最 original 的感覺。這裡沒有 Starbucks，也沒有 7-11。有的就是幾間感覺像停留在 60、70 年代的小館子，一間專門介紹 Kudichin 文化的小型博物館，和一間頗爲殘舊的教堂 Santa Cruz Church。

就讓它繼續沉沉睡！我希望這樣説不會太過份吧。

Khun Sakulthong 和 她的丈夫
Salulthong and her husband

The Kudichin neighborhood is one of Bangkok's oldest. The Portuguese came from Ayutthaya with the Siamese after the fall of Ayutthaya in 1767 and populated the area, creating a lively Portuguese quarter. It is a place with so much heritage that I want to know more.

The car picked us up from Okura Prestige Bangkok Hotel in Silom, it took us almost 30 minutes to get there. It is a small peaceful area, my cooking students and I walked by houses along an alley to the Baan Kudichin museum. Santa Cruz, one of the oldest churches in Bangkok, built by the Portuguese in 1769 after King Taksin granted them the land, is another landmark. We were all very surprised and amazed to find this Portuguese enclave in the middle of cosmopolitan Bangkok.

We had lunch at Baan Sakulthong, a cozy home-style restaurant that is open only on Saturday and Sunday. Of course, I can't compare the food to the authentic food I've had in Portugal, but the restaurant is a testament to how the Portuguese greatly influenced Thai cuisine. The lineage of the Sakulthong family can be traced back to the Portuguese settlers from Ayutthaya. Today, the restaurant follows the family recipe of Chawee Sakhulthong, the owner's great-grandmother.

I didn't know what to order and wanted to be adventurous, so I asked Khun Sakulthong, the owner, to bring us whatever she recommended. We had traditional appetizers *jeeb tua nok* (bird-shaped Thai dim sum) and chor muang (a purple chewy treat stuffed with savory filling). The restaurant's signature main dish is their version of Portuguese-style *kanom jeen*. A filling noodle dish of minced chicken red curry and coconut cream over rice vermicelli (swapping out the white sauce and potatoes of the traditional *kanom jeen*). Another specialty is *tommafad*, a Thai version of cozido, the Portuguese meat-and-vegetable stew. Sakulthong's version is loaded with spices and served with rice in true Thai fashion.

Kudichin is such a different place in Bangkok. The people, the food, and the heritage. It is a hidden treasure.

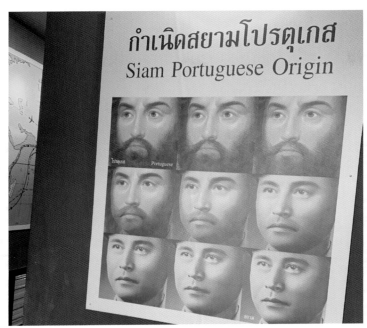

青檸辣椒蒸帶子
Steamed Scallops with Lime and Chilies

當我帶學生和粉絲到曼谷美食遊時，我們到了 Issaya Siamese Club 進餐，其中一個菜式很有趣，以下是我演繹的簡單版本，也是我記憶中最美味，做法也最簡易的蒸帶子。

When I took my cooking students and fans for gastronomy tours in Bangkok, we visited Issaya Siamese Club to try out their dishes. Here I like to share a dish from their interesting menu, with my very simple interpretation. The dish I remember the best, steamed scallops, was also the simplest.

材料（4人份量）		Ingredient (Makes 4 servings)	
刺身級帶子	12隻	Scallops, for sashimi	12
生菜	1棵	Soft lettuce	1 head
新鮮薄荷	2湯匙，切絲	Fresh mint, shredded	2 tablespoons
檸檬葉	6片，切絲	Kaffir lime leaves, shredded	6
指天椒	3隻，切片	Bird's eye chilies, sliced	3
青檸汁	1湯匙	Lime juice	1 tablespoon
鹽	1茶匙	Salt	1 teaspoon

做法 How to cook

預備一隻大碟（要放入蒸鍋時不會貼著鍋底），放上一張牛油紙，再放上帶子。鍋煮滾少許水，放入帶子碟，加蓋蒸至剛熟（應少於2分鐘）。

在每隻帶子放於一塊生菜上面、然後加少許薄荷、檸檬葉和一片辣椒，擠上青檸汁，灑少許鹽即可。

Place the scallops on parchment paper on a plate large enough to fit into the wok without touching the bottom. Place the plate over boiling water in the wok and steam until just cooked. This should take less than 2 minutes.

Place each scallop on a piece of lettuce and add a sprinkle of mint, shred of lime leaves, and a slice of hot chilies. Squeeze over some lime juice and add a pinch of salt to taste.

泰式蒸蛋
Thai-Style Steamed Savory Custard

如果你有機會到訪泰國家庭，這就是他們會用來奉客的一道家常菜，只需幾種材料，做法非常容易。我的泰國朋友會在吃之前再加點魚露，我則喜歡加數滴辣椒油，來點刺激。

This is a very typical home-style dish that you might be lucky enough to savor when visiting a Thai family. It's so easy to do that it takes only a few ingredients. My Thai friend added extra fish sauce before serving, I like to add a few drops of chili oil instead to give an extra kick.

材料（4人份量）		Ingredient (Makes 4 servings)	
雞蛋	3隻	Eggs	3
清水	2湯匙	Water	2 tablespoons
椰奶	2湯匙	Coconut cream	2 tablespoons
魚露	1湯匙	Fish sauce	1 tablespoon
白糖	1茶匙	White sugar	1 teaspoon
蔥	1棵，切片	Spring onion, sliced	1
白胡椒粉	1茶匙	White pepper	1 teaspoon
新鮮芫茜碎	2茶匙	Fresh cilantro, chopped	2 teaspoons
紅辣椒碎	½茶匙（可免）	Red chili, chopped	½ teaspoon (optional)

做法 How to cook

大碗打入雞蛋拂勻，然後加入清水、椰奶、魚露、糖和蔥拌勻。

將混合物倒入耐熱碗或焗碟內。在鑊或鍋中放入一半水煮至微滾，放入蛋以中小火蒸15分鐘至剛熟。

以白胡椒粉調味，灑上芫茜碎和紅椒碎即成。

In a large mixing bowl, whisk eggs then add the water, coconut cream, fish sauce, sugar, and spring onion, stirring to combine.

Pour the mixture into a heatproof bowl or ramekin and steam over gently simmering water in a wok or large saucepan until just cooked, about 15 minutes.

Season with white pepper, cilantro leaves, and red chili.

這樣的一個週末　48 Hours in Bangkok

曼谷跟香港才 2.5 小時的飛機，所以有閒情興致的時候便不妨到曼谷過一個 chillax 的週末。

星期五

在機場 check-in 後，先舒舒服服在貴賓室食一個 late lunch，然後輕輕鬆鬆的上機。

我喜歡在 Silom 區附近 chillout，所以 Okura Prestige Bangkok Hotel 會較適合。而我會入住 club floor 樓層。Why？因為每日都有 evening cocktail 和餐前小食。這樣的 arrangement 對我來說才是 holiday。

食過餐前小食和 evening cocktail，我會到距離酒店一街之隔，位於 Langsuan 區的 Diora Spa 做足底按摩。足底按摩可以令我馬上昏昏欲睡。

星期六

早餐不會吃太多，所以在 club floor 的 continental breakfast 已經心滿意足。一個新鮮出爐的牛角包，再配 blueberry compote，一杯 latte，再加少少西瓜和士多啤梨。滿足！

上午就在酒店的泳池 hea，希望可以看一看我買了但還一直未有時間看的書。要學習這個週末慢活一點。午餐的地方是我近年非常喜歡的 Eat Thai，Eat Thai 位於 Silom 區 Central Embassy 地庫，是一間高檔 food court，最喜歡吃 Eat Thai 的雞飯、魚蛋粉、芒果糯米飯，還有椰子泰式雞蛋仔。吃完午餐我會到位於 Siam Paragon 的 Kinokuniya 書店打書釘，這裡搜羅了很多有關煮食的工具書，有大部份在香港也買不到的，而亮點就是免費包書服務，對「書控一族」簡直是非常貼心的 service。

到 Gourmet Market 逛一逛，買一些食材。對不起，我不是 Big C 粉，so far 只是到過 Big C 一次，沒有買到任何東西。

晚餐？還沒有甚麼計劃，不如再去做足底按摩吧！

星期日

今天早餐多吃一點，這樣便可以 skip lunch，Okura Bangkok 餐廳的西式早餐種類頗多，我尤其喜歡 Egg Benedict，另點多個加了勁多胡椒粉的魚蛋粉，正！！！

早餐後繼續在酒店泳池 chillax。

從酒店坐的士不用 20 分鐘便到機場了，一個愉快又寫意的 weekend 便過去了！

Since Bangkok is only a two-and-a-half-hour flight from Hong Kong, I love to come here on a Friday afternoon and spend a leisurely and chillaxing weekend. Here is how I pamper myself:

Friday

For the afternoon flight, I check in at the airport and have a late lunch at the airport lounge before my flight.

I arrive at Okura Prestige Bangkok and check into my room on the Club floor, with daily complimentary evening canapés and cocktails. I love it.

After my evening cocktail, I walk to Diora Spa located at Langsuan for a two-hour foot massage. I desperately need one and I can easily doze off as soon as the massage starts.

Saturday

I don't eat much for breakfast, so a simple continental breakfast on the Club floor is good enough. All I need is a latte to pick me up, a fruit bowl, a fresh hot croissant with blueberry compote, and I am good to go.

I need to S-L-O-W down. I like to spend my morning at the swimming pool and try to read a few pages of my favorite book (though I need to confess I haven't got the patience and time to read a non-food related book for long).

Then I head to Eat Thai, high-end food court at Central Embassy, for lunch. Chicken rice, boat noodles, and mango sticky rice. And my favorite, *khanom khlok*, which literally means coconut egg waffle.

Afternoon book shopping at Kinokuniya, Paragon Mall. I can spend hours browsing the cook and tool books that I can't usually find in Hong Kong. Yes, and I just love their complimentary book wrapping service.

Next, time for some grocery shopping at Gourmet Market. I get Thai jasmine rice from Chiang Mai, dried herbs, green curry paste for my improvised dish, and Thai chili paste to take home.

My dinner is open, and all depends if I am meeting any friends in Bangkok or not. But I like to end my Saturday evening with another foot massage, adding an extra session for the head and neck.

Sunday

Breakfast at the hotel coffee shop this morning as I will skip lunch: Eggs Benedict sounds great! And a bowl of fish ball noodle soup with extra dash of white pepper.

Should I go to the gym or the pool? Hmmmm… I choose the pool. LOL! And I really want to finish reading at least one chapter of my current book.

Time to leave. Taking a car to the airport on Sunday takes only 20 minutes.

Such a relaxing weekend. It's really not a bad thing to do this once in a while. On the flight, I keep thinking of two Thai dishes I want to cook in the coming week. Live the Thai way!

青胡椒金不換烤鮮魷
Grilled Squid with Green Peppercorns and Basil

這是我很喜歡的魷魚菜式，只需在市場買到新鮮魷魚，餘下的做法很容易，最宜配蔬菜和香飯享用。

This is a dish that I love to make when it comes to squid. It's easy to make as long as you can get some nice fresh squid from the market. Serve with some greens or white rice.

材料（4人份量）		Ingredient (Makes 4 servings)	
魷魚	500克，洗淨剝花，或原隻備用	Squid, cleaned and scored	500g
蠔油	2湯匙	Oyster sauce	2 tablespoons
椰奶	4湯匙	Coconut cream	4 tablespoons
新鮮青胡椒粒	1-2湯匙	Fresh green peppercorns	1 to 2 table spoons
金不換	½杯	Thai basil	½ cup
魚露	1湯匙	Fish sauce	1 tablespoon
青檸汁	2茶匙	Lime juice	2 teaspoons
糖	1茶匙	Sugar	1 teaspoon
新鮮芫茜葉	½杯	Fresh cilantro leaves	½ cup

做法 How to cook

大碗內放入魷魚、蠔油和椰奶拌勻，蓋好放雪櫃醃1小時。

將燒烤平底鑊以中火燒至非常熱，放入魷魚烤至蜷起及不透明（應少於1分鐘）。

將魷魚放入碗內，加入鮮青胡椒粒、金不換、魚露、青檸汁及糖。

拌勻上碟，灑上芫茜即可。

In a large mixing bowl, combine the cleaned squid with the oyster sauce and coconut cream. Cover and marinate in the refrigerator for 1 hour.

Heat a grill pan on the stovetop over medium heat until very hot. Add the squid and cook until it curls and loses translucency. This should take less than 1 minute.

Transfer to mixing bowl and add the fresh peppercorns, basil, fish sauce, lime juice, and sugar.

Toss well and transfer to a serving plate. Garnish with cilantro leaves.

Okura Prestige Bangkok

📞 (+662) 687 9000　　@ www.okurabangkok.com

我愛其商務樓層會員客房，因為會提供晚間 cocktails 和 canapés、免費洗衣服、及每天在 club lounge 的下午茶。早餐可選自助餐或日式早餐，酒店的員工，不論是管家或救生員都很親切友善。

如果你也喜歡「晚間 cocktails 和 canapés」這種旅遊模式，也可考慮曼谷香格里拉的 Krungthep Wing。它更有只供 Krungthep Wing 客人使用的私人泳池。

I love their Club rooms because I can enjoy other added benefits like evening cocktails and canapés, free laundry and pressing, and complimentary daily afternoon tea in the lounge.

If you also fancy the "afternoon tea, evening canapés" travel lifestyle in Bangkok, you can consider the exclusive Krungthep Wing at the Shangri-la. There's even a private swimming pool only for guests of the Krungthep Wing.

The Siam Hotel

📞 (+662) 206 6999　　@ www.thesiamhotel.com

The Siam Hotel 是個隱藏的寶藏，客房少於 60 間，提供超越 5 星級的服務。我入住過很多次，最愛就是住泳池獨立房，有自己的私人泳池，高度私隱又寧靜。

It is simply a hidden treasure. The hotel itself is very exclusive as there are fewer than 60 guestrooms, and the service is really beyond five-star. I have stayed many times — my favorite room is their pool villa which comes with your own private swimming pool.

Eat Thai, Central Embassy

📞 (+662) 160 5995　　@ www.centralembassy.com/store/eathai/

我最愛的菜式：船麵、雞飯、芒果糯米飯、椰子布甸。

My favorite dishes: boat noodles, chicken rice, mango sticky rice, Thai's coconut egg waffle

Baan Sakulthong（Kudichin 料理）

📞 (+662) 605 5665　　🏠 219 Soi Kudee Jeen 3, Bangkok

這餐廳必須提前預約，而且只在週末營業。

The restaurant does not take walk-in patrons. Remember to call and make advance reservation. It is open on weekends only.

Diora Spa at Langsuan 按摩

📞 (+662) 652 1112　　@ www.dioralangsuan.com

食雞專家　Chicken Connoisseur

太興奮了！昨天晚上開始下大雪，今天氣溫只有零下 11 度，我一見到雪就會變得非常 super hyper。哈哈哈！

今次到韓國首爾請了一名私人導遊 Mr Young，託他為我安排了飲食文化的導賞遊，希望可以了解更多韓國的飲食文化，從中給自己一些料理創作的靈感。

Mr Young 昨天帶我去到一間位於江南區附近較為地道的小館子吃 Bulgogi（韓式燒烤），而這裡的韓式燒烤就只有豬肉，因為要在韓國吃上等的牛肉是頗為昂貴的，而小館子的老闆板由於跟 Mr Young 非常相熟，她還親自教授我如何做一個非常地道用來解酒的菜式。

Mr Young 也特別給我介紹了漢城（即首爾之前的名稱）的歷史，使我也明白了韓國跟儒家思想的關係，同時也明白了為何韓國會出現這麼多漢字。

無雞不歡的我又豈能錯過韓式炸雞？我想在首爾的 6 天，我一共吃了 3 次韓式炸雞，又專誠叫了 Park Hyatt 的 Concierge 特別到酒店隔鄰的店子買外賣帶給我（因為下大雪的關係，酒店是那麼體貼，主動找 hotel staff 幫我去買 Take Away）。

Mr Young 告訴我，在韓國吃炸雞已經成為一個新興的飲食文化。韓式炸雞在 60 年代初才出現，越來越受歡迎，更有大學舉辦「嚐雞課程」，考取韓國政府承認的「嚐雞達人」認証（韓文稱為 Chimuli）。

食雞食到有認証，真係好誇張！

I am at the Park Hyatt Seoul. It is minus 11°C. It is snowing.

Mr. Young, my private tour guide, continues with his food and culture tour today. Yesterday, he took me to his favorite bulgogi restaurant, where I had the most juicy and delicious pork so far. Pork is a common protein in Korean cooking, used to flavor and enhance soups, stews, and even some vegetable dishes. Koreans have a special affinity for the belly, shoulder, and neck meat—Korean ancestors believed these are the best parts of the pig and yield the most flavor. The pork-heavy Hangover Stew that the restaurant owner taught me to make is so loving and endearing.

We spend the afternoon visiting the different gates that used to protect the old city of Seoul. And old Seoul is where everything began, including the long history of Korean culinary and etiquette. Mr. Young told me the slim steel chopsticks (*jeotgarak*) and the spoon (*sutgarak*) must be used in parallel, putting rice and food on the spoon with chopsticks, and then eating from the spoon. This is certainly different from the way we use chopsticks in China and Japan.

Korean fried chicken (*peu-ra-i-deu-chi-kin*) only started in the 1960s, as chicken was prepared in the traditional way before that. As fried chicken became more and more popular, different flavors of fried chicken have been introduced and now you can easily find fried chicken dishes in almost every Korean restaurant, including bar.

The love of fried chicken is so strong that companies introduced the concept of a *chimueli* or chicken connoisseur. People attend classes to learn about the trend of fried chicken in Korean culinary history, flavors, and even tasting, and then take an examination to become a *chimueli* or qualified chicken connoisseur. It is not easy. I wonder how many pieces of fried chicken you need to eat before reaching that level?

鹹蝦燦韓式炸雞
Rocha's Style Korean Fried Chicken

傳統的韓式炸雞是將全雞斬件炸好，配醃蘿蔔和各式沾醬。但我改用雞槌，因為我愛拿著雞槌啃！食譜中的香蒜醬油汁會令炸雞滋味又黏手，真是妙絕！傳統上會將雞炸兩次，以確保皮夠香脆。

With traditional Korean fried chicken, the whole chicken is deep-fried and you are served pieces of chopped chicken along with pickled daikon and various dipping sauces. But I wanted to do it differently and make it fun to eat by using all drumsticks—just because I like holding a drumstick and gnawing at it! My special roasted garlic soy glaze will turn your drumsticks yummy and sticky. This is heaven! Traditionally, the chicken is fried two times to ensure that the chicken is crispy enough.

材料（4人份量）

米糠油	750毫升（油炸用）
連皮雞槌	12隻

香蒜醬油汁

雞湯	250毫升
豉油	240毫升
日本或韓國味醂	240毫升
焗蒜泥	300克
黃糖	200克
蔥碎	2湯匙（裝飾用）
烘香芝麻	1湯匙（裝飾用）

Ingredient (Makes 4 servings)

Rice bran oil	about 750ml (for deep-frying)
Chicken drumsticks, skin on	12

Roasted Garlic Soy Glaze

Chicken broth	250ml
Soy sauce	240ml
Japanese or Korean mirin	240ml
Roasted garlic puree	300g
Brown sugar	200g
Spring onion, chopped	2 tablespoons (for garnish)
Roasted sesame seeds	1 tablespoon (for garnish)

做法 How to cook

大湯鍋中倒入米糠油15至20厘米高，以中火加熱至冒煙。將雞槌分批放入炸8分鐘至金黃，盛起放焗盤上備用，炸鍋留用。

香蒜醬油汁：在另一湯鍋中，倒入雞湯、醬油、味醂、焗蒜泥和黃糖，以中火加熱至微微滾，煮20至25分鐘，至醬汁減少⅓，離火備用。

再加熱炸鍋的油，將雞槌分批放入翻炸4至5分鐘至熟透，立刻盛起放入醬汁的鍋中浸3分鐘吸收味道。

將雞槌瀝乾上碟，灑上蔥碎和芝麻，立刻開動。不要害羞，用手吃才是最痛快！

In a large saucepan, heat 15 to 20cm oil over medium heat until you can see smoke coming out. In batches, add the drumsticks and fry for about 8 minutes, until browned. Transfer to a baking sheet and set aside, reserving the oil in the saucepan.

For the glaze: In a separate saucepan, combine the broth, soy sauce, mirin, garlic puree and sugar, bring all to a low boil over a medium heat. Simmer for 20 to 25 minutes, until the sauce has reduced by one-third. Remove from heat and set aside.

Heat the reserved oil in the saucepan until hot. Add the drumsticks and fry again for 4 to 5 minutes in batches, until cooked through. Transfer immediately to the saucepan with the soy glaze and let them sit for about 3 minutes to allow the chicken to absorb the flavors.

Shake the excess glaze off and transfer the drumsticks to a serving plate. Sprinkle with the chopped spring onions and roasted sesame seeds. Serve immediately. Here's the fun part: Eat with your hands and get messy.

焗蒜泥的做法：
將蒜肉放焗碟內，包上錫紙，
放入焗爐以 150°C 焗 1 小時至蒜肉軟身微黃。
將蒜肉放攪拌機中，加少許橄欖油打成幼滑的泥狀，
如太乾可酌量加油。蓋好放雪櫃可保存 2 天。

To make roast garlic puree:
Place peeled garlic cloves in a baking dish
and cover the dish with aluminum foil.
Roast in a 150°C oven for 1 hour,
until the garlic is soft and light golden brown.
Place garlic in a blender along with 2 tablespoons olive oil and
puree to a smooth consistency, adding a little more oil if necessary.
Cover and store in the refrigerator for up to 2 days.

韓式水果酒
Soju Sangria

我愛水果酒（可參考 p.198 我自家版本的葡萄牙水果酒）。我在首爾的柏悦酒店享受了一杯韓國燒酒 Soju，啟發我今次的創作。我以蘆薈汁和水果配韓國燒酒，成為口味清新的飲品，可配我的韓式炸雞，或任何你喜歡的食物。

I love Sangria. (Check for my home version of Portuguese Sangria on p. 198) I was having a drink while staying at Park Hyatt Seoul and got inspired by the famous Korean alcohol soju. I add aloe vera juice and other tropical fruits to make this delightful refreshing drink. It is good to match with my version of Korean fried chicken drumsticks, or simply any food you fancy.

材料（大瓶，6人份量）

無核紅提	80克，切片
韓國梨	50克，切粒
芒果	50克，切粒
布冧	50克，切粒
韓國燒酒	350毫升
蘆薈汁	200毫升
Grand Marnier橙酒	30毫升
蜜糖	1湯匙
青檸	1個，切片後一開四（裝飾用）

Ingredient (Makes 1 jar, enough for 6 servings)

Red seedless grapes, cut in half	80g
Korean pear, diced	50g
Mango, diced	50g
Plums, diced	50g
Soju	350ml
Aloe vera juice	200ml
Grand Marnier	30ml
Honey	1 tablespoon
Lime, cut into slices and then quartered	1 (for garnish)

做法 How to cook

預備一個有蓋的容器，將所有材料放入拌勻（除了青檸片），蓋好放雪櫃冷藏最少4小時。

將混合物轉放入壺中，倒入酒杯內，以青檸片裝飾即成。

In a large container with an airtight lid, combine all the ingredients except the lime. Mix well and cover. Refrigerate for at least 4 hours.

Transfer to a serving pitcher, then pour into wine glasses. Garnish with lime slices and serve.

不要將青檸片放入同浸，因為它的皮帶苦味，我會直接放入酒杯上裝飾。

Do not add the lime too soon as the skin will give out a bitter flavor.
I only add the slices directly to the wine glasses.

住

首爾 Park Hyatt 酒店 （Park Hyatt Seoul）

📞 (+822) 2016 1234　　@ www.hyatt.com

住酒店，其中一種令我苦惱的事莫過於冬天時酒店的中央暖氣不可以自己調校，房裡的空調總是調到最高溫。Park Hyatt 是少數可以自己調校溫度的地方，所以我可在冬天調至喜歡的 20°C，哈。當然服務是一如所料的好，而且房間提供的洗漱用品 是我喜歡的品牌 Aesop。

One thing that I find quite unpleasant is to stay in a hotel that has central heating during winter, but without being able to turn on the AC to its max in the guestroom. Park Hyatt Seoul is a rare hotel that I can really enjoy turning on my in-room AC to 20°C on a cold winter day. LOL. The service is exceptionally good! And I love their in-room AESOP toiletries. One of my favorite skincare brands!

我在首爾三種必吃食物 My three must-eat dishes when in Seoul:

1. 土俗村蔘雞湯（人蔘雞湯）

Tosokchon Samgyetang (Korean ginseng chicken)

🏠 5 Jahamun-ro 5-gil, Chebu-dong, Jongno-gu, Seoul, South Korea

📞 (+822) 737 7444

2. 餃子湯 Dumpling soup

3. 魚餅 Fish cake

簽名會之思前想後　My First Book Signing

這幾星期都在想：我要說甚麼？說一些洋蔥的說話？說一些 juicy stories ？爆料？哎呀，如果我哭了怎麼辦呀？（註：我非常眼淺的。）

昨天晚上睡在牀上一直想著這個簽名會。

倦了，閉上眼，醒了。Today is the day ！

我的首場簽名會就在 Bma Home & Kitchen 舉行。一切給自己做的心理準備、一切為自己構思的情景……沒有出現啊！哈哈！我非常的鎮定，沒有一點緊張的情緒，淡淡定定，在極度 relax 的情況下完成了這個烹飪示範和簽名環節。現在一直寫著這篇日誌，一直看著今天同事們給我拍的照片，眼睛才有淚光啊。

是的，12 個月前我還是 1 名寂寂無名的料理導師，還記得在最初入行的感覺和情景。今天看見一班同學在這個週末來到這場示範班，旨不在嚐嚐我做的新菜式「蟹肉沙律釀炸薯皮」，而是給我的支持。看到同學們排住隊給我簽名，我聽到你們的歡樂聲、給我的掌聲，實在是感動的。

有人說熱鬧過後是寂靜的。倒不是啊！如果是遺憾，我想會是在離開 Bma 的時候，走在長長的走廊，突然想起 Molly 於幾星期前跟我透露關門的意向。要經營一個料理教室實在不容易！難道這是我最後一次在 Bma 辦料理班？

後記：2019 年 3 月 Bma 終於關門了。我一直以來視 Molly 和 Bma 為我的貴人，也是透過他們我才認識到不同的贊助商，他們每一位同事都是有 heart 有 passion，能夠認識到他們實在是我的福氣。

I had so many different thoughts before my book signing at Bma Home & Kitchen. I was thinking how to start? What if I cried ? What should I wear?

Today is the day.

I was extremely calm during the session. I made a little speech and those were my real thoughts. Good, I shed no tears. I managed to thank Molly (the owner of Bma Home & Kitchen) for inviting me to join her studio almost a year ago, when I was still hardly known by anyone, and for not even hesitating for a second when I asked her to host the book signing.

Early fans from Bma Home & Kitchen were there and we had a very fun and loving cooking demo. I invited the students to join me in plating the Avo Chicken (p. 73 of *The Ham Har Chaan Cookbook*) to upgrade the whole presentation, and I unveiled a new dish, Crabmeat Salad in Fried Potato Skins. They loved it!

I was moved when people started lining up for me to sign the book. The most touching part was seeing many of the fans who had supported me from the very beginning of my teaching career; a simple hug meant a lot to me.

While this is a day filled with love and laughter, I also feel sad because I know it might be the last time for me to host a cooking session at Molly's studio. Molly told me a few weeks ago that she may have to close down the business as the operating costs are simply too high. It's not easy running your own business these days.

Addendum: As Molly had feared, Bma Home & Kitchen closed its door in March 2019. And that book signing was the last time I had a cooking class with Molly and her team. Molly, I thank you for everything.

蟹肉沙律釀炸薯皮
Crabmeat Salad in Fried Potato Skins

記得太古廣場的 Dan Ryan's 嗎？我超愛它的酸忌廉芝士焗薯皮，它亦是我這道菜的靈感來源。我用新薯代替焗薯，釀好的薯仔有點像蘑菇，實在太可愛了。

Do you remember Dan Ryan's Chicago Grill at Pacific Place? I just love their "potato skins with sour cream and cheese." That's the inspiration behind these miniature versions.

材料（4人份量）

新薯	6粒（越小越好）
米糠油	適量（炸薯用）
蟹肉	150克
日式蛋黃醬	3湯匙
蔥花	1茶匙
鹽和胡椒粉	適量
檸檬皮蓉	2茶匙

Ingredient (Makes 4 servings)

New potatoes (the smaller the better)	6
Rice bran oil	for deep-frying
Crabmeat	150g
Japanese mayonnaise	3 tablespoons
Spring onion, chopped	1 teaspoon
Salt and pepper	to taste
Grated lemon zest	2 teaspoons

做法 How to cook

在焗爐中層放上一個焗架，預熱至180°C。

用利刀在新薯上刺數下，然後將全部新薯用錫紙包好，再放在焗盤上，入焗爐焗40分鐘至軟身。取出薯仔，在每個薯仔的頂部削去數毫米。然後用茶匙小心刮出薯仔肉，放到大碗中備用。

在大湯鍋中倒入約8厘米（3吋）高的油，以中火加熱，放入薯皮炸2至3分鐘至金黃，期間反轉一次，取出備用。

將刮出的薯肉壓成蓉，加入蟹肉、蛋黃醬和蔥花拌勻，加鹽和胡椒粉調味，將薯蓉填入炸薯皮內，灑上檸檬皮蓉即可享用。

Position a rack in the center of the oven and preheat the oven to 180°C.

Stab the potatoes a few times with a sharp knife, then wrap them all up in a piece of foil. Place on a baking sheet and roast for 40 minutes, or until soft. Cut off less than 1cm from the top of each potato, carefully scoop the guts into a large bowl, and set aside.

Heat about 8cm (3 inches) oil in a large saucepan over medium heat. Add the potato skins and fry, turning once, for 2 to 3 minutes, until browned. Set aside.

Mash the potato flesh well. Stir in the crabmeat, mayonnaise, and spring onion. Season with salt and pepper to taste. Mound the mixture into the potato skins and sprinkle with the lemon zest.

紅白歌唱大賽 · Red and White

　　大家有沒有看過《紅白歌唱大賽》？80 後、90 後的朋友一定不會看懂這篇日誌，但如果你像我出生於 1971 年或以前、在香港或曾經在香港居住過，大家是否記起在某年的 12 月 31 日，TVB 總會隆重其事人造衛星直播由日本 NHK 電視台製作的《紅白歌唱大賽》？

　　嗯！嗯！我記得翁倩玉姐姐那件白色蕾絲半透明加頭上白色錦花的招牌戰衣，還有那個招牌動作；一些我當年完全不懂得 appreciate，穿著日式和服未唱先哭到「妝都甩埋」的演歌歌手；然後接近 12 點家人就會把電視轉台到 TVB 明珠台，觀看於銅鑼灣一年一度遊艇會附近舉行的除夕新年鳴放禮炮儀式直播，媽媽和嫲嫲就會準備好香檳，小朋友喝的當然是 fruit punch 啦。

　　而嫲嫲總愛想當年，告訴我和妹妹，爺爺在生前帶她到銅鑼灣出席鳴炮儀式，然後一起喝著香檳、跳著著舞。這個「話當年」都是我們其中的一個節目。

　　由於爸爸剛在 11 月才離開了我們，媽媽已經吩咐了今年將不會有任何的聖誕和新年慶祝活動。所以今天晚上便靜靜的、簡單的過一個除夕夜。

　　吃完晚飯，突然想起小時候的除夕夜，想起紅白歌唱大賽。打開很久已經沒有看過的 TVB，原來這個年頭已經再沒有轉播紅白大賽了。想一想，原來自己也改變了「看電視」的習慣和模式，我喜歡的是 on-demand，不再是 live broadcast（除了新聞節目）。我打開 YouTtube，找到翁倩玉姐姐當年在紅白大賽唱的《愛的迷戀》視頻，再一次看見那個招牌動作。給自己倒了杯 Pinot Noir，突然想起港式西餐的「紅湯」和「白湯」。看看電話，距離 2019 年還有 2 小時。走到廚房，打開雪櫃，突然要做紅白兩款湯來喝啊！

　　後記：根據 Wikipedia 的解釋，「演歌」是日本頗具代表性的傳統歌曲類別，演唱風格反映出日本人的慣有情感表達方式。演歌也被稱爲艷歌和怨歌。

When I was young, I spent New Year's Eve at home with my parents, family, and friends, enjoying a big feast with plates and plates of food, and everyone holding a glass of champagne during the countdown to the New Year. (Well, soda water mixed with cubes of peach and apple for the children.) On the television was a show from NHK television network called *Kohaku Uta Gassen*, a singing competition. (The program is still showing in Japan now after 65 years!)

My family no longer has gatherings like that, and this year Mom decided, since Daddy left us not long ago, that we should show our respect over the holidays, so everyone in the family canceled any Christmas and New Year celebrations. For me, I had a very simple dinner at home, and turned on the television to search for a movie. That's when I remembered *Kohaku Uta Gassen*; but I was quite disappointed to find that the local television network is no longer broadcasting the program.

But it got me thinking: I remembered that the name of the program translated as "Red and White Singing Contest," with red representing the female singers and white representing the male singers. How cliché… hahaha.

But "red" and "white" reminded me of the local borscht soup and cream of chicken soup that I love. The soups are not made in the traditional way but they are two of my favorite fancy Hong Kong – style Western dishes.

I checked the fridge and immediately headed to the supermarket to buy some missing ingredients. As I am writing this down, both soups are simmering. This is my personal 2018 New Year's Eve countdown. And I know 2019 is going to be a very nice one!

鹹蝦燦羅宋湯（紅湯）
Rocha's Style Borscht Soup

在東歐，羅宋湯非常普遍。但羅宋湯的英文 borscht，其實也指烏克蘭地區一些用紅菜頭做成的紅色湯。在香港喝到的羅宋湯都不是正宗的，因為都不含紅菜頭，湯的紅色是來自茄膏。下次你煮羅宋湯時，可以試試加一罐紅菜頭，即時增添原著味道。

Although borscht is a soup common throughout Eastern Europe, in English the word 'borscht' is associated with the soup's Ukrainian origin, made with beetroots, which give the dish its distinctive red color. The borscht that I have been eating in Hong Kong is not the traditional and authentic one as it does not contain beetroot. But it does look the same with its red color from tomato paste. You may like to add a can of beetroots to give it an original touch.

材料（4人份量）		Ingredient (Makes 4 servings)	
洋蔥	1個，切片	White onion, sliced	1
紅洋蔥	1個，切片	Red onion, sliced	1
橄欖油	2湯匙	Olive oil	2 tablespoons
紅蘿蔔	80克，切片	Carrot, sliced	80g
西芹	80克，切片	Celery, sliced	80g
番茄	4個，連皮一開四	Tomatoes, skin-on, quartered	4
茄膏	1½湯匙	Tomato paste	1½ tablespoons
糖	1茶匙	Sugar	1 teaspoon
雞湯	750毫升	Chicken broth	750ml
清水	500毫升	Water	500ml
牛肉粒（適合燉煮的部位）	100克	Beef cubes for stew	100g
香葉	3片	Bay leaves	3
圓椰菜	½個，切條	Round cabbage, sliced	½
檸檬汁	1湯匙	Lemon juice	1 tablespoon
鹽和胡椒粉	適量	Salt and pepper	to taste

做法 How to cook

在一個大湯鍋中下橄欖油，以中火炒洋蔥至軟身，約5至8分鐘。加入紅蘿蔔、西芹和番茄，繼續炒4分鐘。

放入茄膏和糖炒1分鐘。然後放入雞湯、水、牛肉粒和香葉，蓋好以中火煮30分鐘，期間不時攪拌，煮至牛肉軟腍熟透。

加入椰菜，以小火再煮30分鐘至椰菜軟身。離火，加入檸檬汁，以鹽和胡椒粉調味，棄掉香葉即可享用。

In a large saucepan, sauté the onions in the olive oil over medium heat until soft, 5 to 8 minutes. Add the carrot, celery, and tomatoes and continue to sauté for another 4 minutes.

Stir in the tomato paste and sugar and sauté for 1 minute. Add the broth, water, beef cubes and bay leaves, Cover and cook over medium heat, stirring occasionally for about 30 minutes. until the beef cubes are tender and cooked.

Add the cabbage and simmer over low heat for another 30 minutes, until the cabbage is soft. Turn off the heat, add the lemon juice, and season to taste. Discard the bay leaves and serve.

我喜歡加數滴 Tabasco 或 piri piri 辣汁，加點刺激。
如果你想「吃得」更豐富，可以在湯煮好前 15 分鐘，
加入芽菜同煮，即時帶點中國風味。

I like to add drops of Tabasco or piri piri to give the soup an extra kick.
If you love to have more substance to the soup, you may want to add
bean sprouts just 15 minutes before serving. This is what my mom
likes to do to give a "Chinese" touch to the soup.

鹹蝦燦忌廉雞湯（白湯）
Rocha's Style Cream of Chicken Soup

我以前喜歡開一罐忌廉雞湯，然後拌飯吃。有時去到連鎖快餐店，我也會叫「白湯」，但在湯中很難找到一塊雞肉。不過近年我更喜歡自己做忌廉雞湯，很治癒呢！

I've always loved making myself cream of chicken soup from a can and serving it with rice! And when I visit local fast food chain stores, I still sometimes fancy a bowl of "white soup" (although you can hardly find a piece of chicken meat in that soup). But in recent years, I prefer to make my own homemade version, as it's more comforting and therapeutic.

材料（2人份量）

烤雞腿肉	60克，切絲
洋蔥	30克，切粒
橄欖油	2湯匙
無鹽牛油	1塊
麵粉	1湯匙
全脂奶	300毫升
雞湯	500毫升
忌廉	100毫升
紅甜椒	60克，切粒
鹽和胡椒粉	適量

Ingredient (Makes 2 servings)

Roasted chicken thighs, shredded	60g
White onion, diced	30g
Olive oil	2 tablespoons
Unsalted butter	1 knob
Plain white flour	1 tablespoon
Full-cream milk	300ml
Chicken broth	500ml
Cooking cream	100ml
Red bell pepper, diced	60g
Salt and pepper	to taste

做法 How to cook

將雞肉切絲或像我這樣的用手拆肉。我喜歡雞腿較軟，但你也可以用雞胸，也可隨喜好決定是否去皮，我自己則喜歡雞皮的質感。

在大湯鍋中加入橄欖油，以中火炒洋蔥5至8分鐘，轉小火，放入牛油和麵粉，不斷推炒20秒，然後加雞肉，倒入牛奶和雞湯。煮至滾後以中火再煮10分鐘，至湯轉奶白色。

加入忌廉和紅椒粒，開蓋煮2分鐘至紅椒稍軟身。離火，以鹽和胡椒粉調味即成。

Shred the meat from the chicken. I like chicken thighs here as the meat is tender, but you could also use breasts. You can shred with or without the skin. Personally, I like skin-on as I love the texture of chicken skin.

In a large saucepan, sauté the onion in the olive oil over medium heat until soft, 5 to 8 minutes. Turn the heat to low, drop the knob of butter into the saucepan, followed by flour. Stir gently for 20 seconds, then add the chicken, milk and broth. Bring to a boil and simmer over medium heat for 10 minutes, until the soup turns milky.

Add the cream and diced bell pepper and simmer, uncovered, for 2 minutes, until the bell pepper turns slightly soft. Remove from the heat and season to taste with salt and pepper.

我最喜歡以白飯配忌廉雞湯，但你也可以餅乾或麵包粒伴食。
這食譜方便快捷，在 30 分鐘以內就可以完成。

I love cream of chicken soup with white rice!
Or you may like to add crumbled cream crackers or croutons.

This is a quick and easy recipe. It should not take you
more than 30 minutes start to finish. Enjoy it!

第二部　Part Two

再煮

More Cooking

不能自滿，毋忘初衷
尋食、尋突破
如何是好
煮
煮我喜歡的
快樂是簡簡單單。

If you are always trying to be normal,
you will never know how amazing you can be.
~Maya Angelou~

新年快樂 · New Year's Wishes

今天是 2019 年的第 1 天。我要寫下今年要做的事情：

我要出版第二本書。

我要自我增值，尤其在飲食文化和歷史方面。

我要煮更多新菜式。

我要家人開心。

我要家裡 3 個毛孩健健康康。

我要每天活得開開心心。

我要今天晚上在夢裡見到毛孩「大哥哥」。他於 2018 年 1 月 1 日離開了這世界。

Chico 現在已經成為 3 個毛孩中的首領。
Chico is the leader of the pack now.

My wishes for the new year:

I want to write another book.

I want to learn more.

I want to cook more new dishes.

I want my family to be happy.

I want my three dogs to be healthy.

I want to feel good and happy every day.

I want to see my dog in a dream tonight. He left me on January 1, 2018.

蒜香脆鱈魚
Grilled Cod with Garlic Panko

我很喜歡吃炸魚薯條（Fish and Chips）。最近在家附近一間餐廳吃午餐時，突然想到可用麵包糠做炸魚，代替一般的粉漿。為了增添一點亞洲風味，我選用了日本麵包糠，效果很好呢！

I love fish and chips. I was recently enjoying a plate for lunch at a restaurant near my home, and suddenly had the idea of replacing the usual fish-and-chips batter with bread crumbs. And to give it an Asian touch, I thought about using Japanese panko crumbs.

材料（4人份量）

日式麵包糠	120克或以上
橄欖油	3湯匙
檸檬汁	3湯匙（1個份量）
蒜頭	2瓣，拍碎
新鮮番茜碎	2湯匙
乾辣椒碎	1茶匙
鱈魚柳	4塊，每塊約150克
鹽和胡椒粉	適量

Ingredient (Makes 4 servings)

Japanese panko	120g or more
Olive oil	3 tablespoons
Lemon juice	3 tablespoons (from 1 lemon)
Garlic cloves, crushed	2
Fresh parsley, chopped	2 tablespoons
Dried chili flakes	1 teaspoon
Cod fillets	4 (about 150g each)
Salt and pepper	to taste

做法 How to cook

以中火燒熱一個坑紋煎鑊。將日式麵包糠、橄欖油、檸檬汁、蒜頭、番茜和乾辣椒放中型碗中拌勻，如果混合物太濕，可多加一點麵包糠。

鱈魚以鹽和胡椒粉調味，然後將兩面沾滿麵包糠，用手輕輕按實。

煎鑊要燒得非常熱，然後放入鱈魚，每面煎約3分鐘至金黃香脆。上碟即可。

Preheat a griddle pan over medium heat. Mix together the panko, olive oil, lemon juice, garlic, parsley, and chili flakes in a medium bowl. If the mixture is too wet, add more panko.

Season the cod with salt and pepper and dip each side into the panko mixture. Press down with your hands to ensure the panko coats the fillets evenly.

Make sure the griddle is very hot. Place the fillets on the hot griddle and cook for about 3 minutes on each side, until browned and crispy all over. Cut each fillet in half and serve.

香焗芝士羊肉馬鈴薯

Baked Lamb with Potatoes and Parmesan

數年前在香港，我在姨母家中過新年，她煮了一個羊肉，如牛油般軟腍，馬鈴薯的表面焗得脆脆，中間的部份則入口溶化，令我回味至今。

I remember having this amazing dish a few years ago in my auntie's home in Hong Kong. The meat is soft as butter, and the potatoes get crispy on top but stay soft through the layers.

材料（4人份量）		Ingredient (Enough for a feast)	
橄欖油	4湯匙	Olive oil	4 tablespoons
焗薯	6個，去皮，切½厘米片	Baked potatoes, peeled and cut into ½cm-thick slices	6
鹽和胡椒粉	適量	Salt and pepper	to taste
羊腿肉	1公斤，切1-2厘米方粒	Leg of lamb, boneless, trimmed and cut into 1-2cm cubes	1kg
車厘茄	500克，切半	Cherry tomatoes, halved	500g
新鮮迷迭香碎	2湯匙	Fresh rosemary, chopped	2 tablespoons
蒜頭	4瓣，切碎	Garlic, chopped	4 cloves
巴馬臣芝士碎	100克	Parmesan cheese, grated	100g

做法 How to cook

預熱焗爐至180°C，預備一個鑄鐵鍋或不黏的燉鍋，倒入2湯匙橄欖油，先將部份的馬鈴薯片鋪滿鍋底。然後放入羊肉粒，再加調味。

將車厘茄在洗濯盤上用手搾去籽，然後放在羊肉粒上，灑上一半迷迭香和一半蒜頭，鋪上芝士。

鋪上餘下的薯片、車厘茄、迷迭香、蒜頭及調味，灑上芝士及澆上2湯匙橄欖油。

蓋上鍋蓋或用錫紙包好，放入焗爐焗35分鐘。然後開蓋或撕去錫紙多焗50分鐘，至面層金黃香脆，出爐後靜置10分鐘，即可享用。

Preheat the oven to 180°C. Pour 2 tablespoons oil into a cast iron Dutch oven or oven-proof nonstick casserole. Arrange some potato slices over the bottom. Scatter over meat cubes and season with salt and pepper.

Squeeze the halved tomatoes over the sink to get rid of the seeds, then scatter over the lamb, along with half the rosemary and half the garlic. Sprinkle with grated cheese.

Layer the remaining potatoes, tomatoes, rosemary, garlic, seasoning and remaining cheese. Drizzle the remaining olive oil all over.

Cover the pot or casserole with the lid or foil, transfer to the oven, and bake for 35 minutes. Uncover and continue to bake for another 50 minutes, until the top is golden and crusty. Let rest for 10 minutes before serving.

雜錦燉肉伴饅頭
Party Meat Stew with Chinese Plain Buns

這是我另外一款的自創食譜，靈感來自辣椒燉豆，是一道非常濃郁的燉肉鍋，伴饅頭享用。用饅頭而非白飯和墨西哥玉米片，好像有點奇怪，但你一定會愛上它的。

This is my own recipe inspired by chili beans: a very rich and tasty meat stew served with Chinese plain buns. Chinese buns instead of white rice or nacho is kind of crazy, but trust me, it's an innovative way to enjoy meat stew. Try it and you will fall for it. Enjoy!

材料（6人份量）		Ingredient (Makes 6 servings)	
初榨橄欖油	1湯匙	Extra virgin olive oil	1 tablespoon
洋蔥	1個，切碎	White onion, finely chopped	1
新鮮迷迭香碎	2湯匙	Fresh rosemary, chopped	2 tablespoons
牛肉粒（適合燉煮的部位）	350克	Beef cubes for stew	350g
豬肋骨	8條，約200克，切細件	Pork ribs, cut into small pieces	8 (200g)
雞肉腸或豬肉腸	4條，約150克	Sausages (chicken or pork)	4 (150g)
罐頭原粒番茄	2罐，連汁共400克	Plum tomatoes (canned)	2 (400g) cans, including liquid
茄膏	2湯匙	Tomato paste	2 tablespoons
熱水或雞湯	150毫升	Hot water or chicken broth	150ml
新鮮羅勒葉	5塊	Fresh basil leaves	5
鹽和胡椒粉	適量	Salt and pepper	to taste
饅頭	6個	Chinese plain buns	6

做法 How to cook

大湯鍋以中火加熱，放入洋蔥和迷迭香不斷拌炒5分鐘至軟身。加牛肉、排骨和香腸煮5分鐘。

加入番茄連湯汁、茄膏、雞湯及羅勒葉，拌勻煮滾後調至小火，以鹽和胡椒粉調味，然後蓋上蓋子，煮1至1½小時至湯汁濃稠有光澤及肉身軟腍。煮的期間要每15分鐘檢視一下，如煮乾了可酌量加熱水。趁熱伴饅頭享用。

Heat the oil in a large saucepan over medium heat and sauté the onion with the rosemary for 5 minutes, until softened, stirring occasionally. Add the beef, ribs and sausages and continue to cook for 5 minutes.

Stir in the tomatoes with liquid, the tomato paste, broth and basil. Bring to a boil and immediately lower the heat to a simmer. Season with salt and pepper, cover, and cook gently for 1 to 1½ hours, until the sauce is thick and shiny and the beef and ribs are tender. It is important to check the pot every 15 minutes or so and add more hot water if the stew gets dry. Serve hot accompanied with Chinese plain buns.

簽名會 · Book Signing

此時、此刻。感謝您們！

A million thanks !

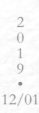
吃甜點　Bad Days

不開心的時候，你在做甚麼？

有人說過吃甜品會令身體產生一些化學變化，大腦會釋放一些「正能量」。

好吧！今天我要做一些甜點來吃，吃完後必定會輕鬆過來。

如果有一天你看到這篇日誌：放鬆吧！給自己找一個亮點，凡事都有兩面性。沒有嚐過不快樂，又怎會知道快樂的感覺是怎樣？

Life is full of ups and downs. What do you do when you are feeling down or grumpy? People often said sugar can boost our energy and lift our mood. There are days when I don't feel right, or when I feel I need a break. I simply make myself a sweet treat to cheer me up. It works most of the time.

Find yourself a cocoon, and things will be okay.

攝於欣澳巴士站
Taken at Sunny Bay bus stop

木瓜青檸香草乳酪
Papaya and Lime with Vanilla Yogurt

這乳酪充滿果香，人見人愛。這甜品很簡單，隨手就預備好。不開心的時候，甚麼也別多想，深呼吸一下，在安樂窩中享用這杯乳酪，給身心放鬆一下。

This fruity mood enhancer is so easy and handy that everyone can enjoy it. What I love about this recipe is that you don't need to think too much. And it's so comforting and just what you need on days when you don't feel quite right. Just take a deep breath, don't think, and get into your cocoon.

材料（4人份量）

木瓜	1公斤
青檸汁	1湯匙
糖霜	2湯匙
低脂乳酪	350毫升
雲呢拿油	1茶匙

Ingredient (Makes 4 servings)

Papaya	1kg
Lime juice	1 tablespoon
Icing sugar	2 tablespoons
Low-fat yogurt	350ml
Vanilla extract	1 teaspoon

做法 How to cook

木瓜去皮去核，放在碗中用叉壓爛，記得不要用攪拌機，會失去口感變成木瓜奶昔。

加入糖和青檸汁輕手拌勻，再輕手拌入乳酪和雲呢拿油，分別放入4個玻璃杯中。

用保鮮紙蓋好，冷藏後隨時享用，放雪櫃可保存2天。

Peel the papaya, remove the seeds, and place in a bowl. Using a fork, mash the flesh until smooth. Please don't use a food processor as you simply do not want this to turn into a papaya smoothie. LOL!

Add the lime juice and sugar and gently mix everything together. Fold in the yogurt and vanilla. Spoon the mixture into 4 glasses.

Cover with plastic wrap and place in the fridge until ready to serve, up to 2 days. Or simply take a spoon and start digging into cheer you up.

葡式朱古力米布甸
Chocolate Arroz Doce

這是我個人版本的葡式米布甸，加入可可粉。朱古力和椰子非常合拍呢！傳統做法可參考《鹹蝦燦之味》第 235 頁，你也可以按口味調整糖的用量。

This is my personal chocolate interpretation of Portuguese rice pudding, arroz doce (see the traditional recipe on p. 235 of *The Ham Har Chaan Cookbook*). You can adjust the sweetness level according to your own preference. Coconut and chocolate... they are absolutely on very good terms with one another!

材料（2人份量）

全脂奶	500毫升
糖粉/幼砂糖	30克
可可粉	2茶匙
Arborio意大利米	70克
雲呢拿油	½茶匙
紅桑莓	40克

Ingredient (Makes 2 servings)

Full-cream milk	500ml
Caster sugar	30g
Cocoa powder	2 teaspoons
Arborio risotto rice	70g
Vanilla extract	½ teaspoon
Raspberries	40g

做法 How to cook

中型鍋中放入牛奶、糖和可可粉，以小火加熱，攪拌至糖溶化。加入米略拌，以中火煮滾後調至小火，拌煮30分鐘至濃稠杰身，米軟身煮透。加入雲呢拿油和紅桑莓拌勻，離火待1分鐘。

將米布甸舀入喜歡的杯中，即時享用。

Combine the milk, sugar, and cocoa powder in a medium saucepan and stir over low heat until the sugar has dissolved. Add the rice and stir briefly. Bring to a boil and reduce the heat to low. Cook for 30 minutes, stirring occasionally, until it looks all thick and creamy and the rice is tender and fully cooked. Stir in the vanilla and raspberries. Remove the pan from the heat and set aside to rest for 1 minute.

Spoon the rice mixture into your favorite cups. Serve immediately.

我的料理班被 KO　　Cancelled!

料理班間中都會遇上「收生人數不足」而給 KO 的情況。每次知道自己的料理班給 KO，我的心情是不好受的。我會問自己是否菜式不吸引？又或是其他原因，例如學校考試？阿 Sir 教得唔好？

Towngas 推出了新料理班課程，我就是好似「睇住個 stock market 咁」不停 keep track 住報班的顏色由綠色（接受報名）到黃色（尚餘少量餘額），及我最喜歡見到的紅色（滿額）。由綠變紅，這狀況好似已經很久沒有再出現了！慘！慘！慘！這就是我的壓力啊！

早在半年前，已經知道會在不久將來出現的樽頸問題，於是便開始尋找「化解」方法。最後還是給自己的鹹蝦燦品牌做一個 re-positioning，旨在吸納更多 fans、更多同學。有人氣 = 有價值，有價值 = 有 say，這句說話好 cliche，但現實就是這樣。

看看被 KO 料理班的菜式，真的不俗啊！朋友們之前有嚐過也讚不絕口。沒有來上料理班的同學，你們走寶了！哈哈哈！

Oh no! A scary thing happened. My class for this week has been cancelled. Reason: Not enough enrollment.

As soon as one of my cooking classes is posted, I keep a very close watch to check the status, and I have to admit that I love seeing the alert going from green (open for registration) to yellow (a few seats available) to finally red (full!). Some instructors feel "it's totally okay" when a class is cancelled, but I think it is not totally okay.

What exactly happened? Are my dishes not attractive? Are the class fees too high? Or maybe I am no longer needed. Well, you can call me Mr. Negative — but it's not a bad idea to think about the worst and start making contingency plans.

So now I know that I need something more interesting and a new breakthrough… Maybe diversify my cooking… Cooking classes are not the only way to go… Definitely go beyond Facebook… Okay, IG… YouTube, yes video!… My own portal… Or travel!

Yes, I need to re-position myself, get more people in my cooking classes, more people following social media. But mostly, I need a change!

When in doubt, or whenever I engage myself into some serious thinking, instead of locking myself up and sitting in front of a computer, I cook. I looked at my own black book (which is my cooking journal with notes on ingredients, measurement changes, next step of action, questions) and found two interesting dishes that I have made before: beef stew and chicken with cabbage. I cooked them and invited my friends over for dinner. Thumbs up!

葡式燉牛肉
Carne Guisada (Beef Stew)

每次做葡式燉牛肉，我都覺得很治癒及充滿創意，因為它沒有既定的副材料，你可以按個人喜好加入西芹、菇菌或南瓜。一樣可以做出美味的菜式。

I always find preparing the Portuguese version of beef stew to be so therapeutic—and creative. This is because there's no absolute yes or no to ingredients other than beef : Feel free to add celery, mushrooms or pumpkin, if you want to.

材料（4人份量）		Ingredient (Makes 4 servings)	
紅酒	100毫升	Red wine	100ml
砵酒	100毫升	Port wine	100ml
香葉	2片	Bay leaves	2
蒜頭	2瓣，切片	Garlic, sliced	2 cloves
鹽和白胡椒粉	適量	Salt and white pepper	to taste
黑胡椒粉	1湯匙	Ground black pepper	1 tablespoon
新鮮番茜碎	1湯匙	Fresh parsley, chopped	1 tablespoon
牛肉粒（適合燉煮的部位）	650克	Beef cubes, for stew	650g
橄欖油	3湯匙	Olive oil	3 tablespoons
洋蔥	2個，切片	White onions, sliced	2
紅蘿蔔	1條，切碎	Carrot, chopped	1
番茄	2個，連皮一開四	Tomatoes (skin-on), quartered	2
馬鈴薯	2個，去皮切粒	Potatoes, peeled and cubed	2
雞湯或牛肉湯	100毫升	Chicken or beef broth	100ml

做法 How to cook

在大碗中，放入紅酒、砵酒、香葉、蒜、鹽、白胡椒粉、黑胡椒粉和番茜拌勻，放入牛肉粒，用手撈勻並按摩一會。蓋好放雪櫃至少6小時或過夜。

將肉和醃汁分開，醃汁留用。將橄欖油倒入大燉鍋，以中火燒熱，放入牛肉粒煎約5分鐘至每邊微黃，盛起放碟上。將洋蔥放入鍋中炒4至6分鐘至軟身。

牛肉粒回鍋，加入醃汁、紅蘿蔔及番茄，蓋上蓋子，以中火煮50至70分鐘。放入薯粒，再蓋好以小火多煮30分鐘，如太乾可加湯，煮至湯汁稍濃稠及減少一半水量，調味。棄去香葉，以白飯或意粉伴食。

In a large mixing bowl, prepare a marinade with the red wine, port wine, bay leaves, garlic, salt, white pepper, ground black pepper, and parsley. Add the beef cubes and give them a good massage. Cover and marinate in the refrigerator for at least 6 hours or overnight.

Drain the meat, reserving the marinade. Heat the olive oil in a large casserole over medium heat. Add the beef and cook, turning the beef to seal all round, until they are lightly browned, about 5 minutes. Transfer the beef to a plate. Add the onions to the pot and cook until softened, about 4-6 minutes.

Return the beef to the pot and add the reserved marinade, carrot, and tomatoes. Cover and cook over gentle heat until the meat is almost tender, 50 to 70 minutes. Add the cubed potatoes, cover, and cook over low heat, adding broth if the stew becomes too dry, for another 30 minutes, until the sauce becomes slightly thick and reduced by half. Season to taste. Discard the bay leaves and serve at once with rice or pasta.

椰菜鷹嘴豆燉雞腿
Chicken Thighs with Cabbage and Chickpeas

這燉雞腿在葡萄牙中部及北部很普遍，簡單易做，最適合一家人吃飯或派對時享用。

This hearty chicken stew is commonly found in central and northern Portugal. Because it's so easy and straight-forward, it's ideal for a family dinner or house-warming party.

材料（4人份量）		Ingredient (Makes 4 servings)	
橄欖油	2湯匙	Olive oil	2 tablespoons
洋蔥	1個，切碎	White onion, chopped	1
煙肉	150克，切粒	Bacon, diced	150g
西班牙腸	80克，切粒	Chorizo, diced	80g
紅蘿蔔	1個，切粒	Carrot, diced	1
蒜頭	3瓣，切碎	Garlic, chopped	3 cloves
香葉	2片	Bay leaves	2
新鮮百里香	2枝	Fresh thyme	2 sprigs
黑胡椒粒	2湯匙	Black peppercorns	2 tablespoons
雞腿	4隻	Chicken thighs	4
清湯或水	150毫升	Broth or water	150ml
圓椰菜	半個，切條	Round cabbage, sliced	½
罐頭鷹嘴豆	150克	Chickpeas (canned)	150g

做法 How to cook

在易潔鑊或深煎鑊中加熱橄欖油，放入洋蔥以小火炒5分鐘至軟身，調至中火，加入煙肉、西班牙腸、紅蘿蔔、蒜、香葉、百里香及胡椒粒，再炒2分鐘。

放入雞腿，加湯或水至剛蓋過雞腿。煮至剛滾立即調至小火，蓋好再煮20分鐘，至湯汁減少一半及雞腿熟透，盛起雞肉保溫備用。

棄去湯汁內的香葉和百里香，放入椰菜及鷹嘴豆，以中火邊煮邊攪拌至混合，轉小火再多煮5至10分鐘，將椰菜煮至你喜歡的質感，將雞腿伴蔬菜及香濃湯汁享用。

Heat the olive oil in a large nonstick sauté pan or frying pan. Add the onion and cook over a low heat, stirring occasionally, for 5 minutes, until softened. Raise the heat to medium and add the bacon, chorizo, carrot, garlic, bay leaves, thyme, and peppercorns. Cook, stirring constantly, for 2 minutes.

Add the chicken and enough broth just to cover. Bring just to the boil, reduce the heat, cover, and simmer gently for 20 minutes, until sauce is reduced by half and chicken is cooked. Remove the chicken and keep warm.

Remove and discard the bay leaves and thyme sprigs from the cooking liquid. Mix the cabbage and chickpeas into the cooking liquid. Cook over medium heat, stirring, until heated through and well blended. Cook over low heat for 5 to 10 minutes longer, until the cabbage is cooked according to your preference. Serve the chicken with the vegetables and rich and flavorful sauce.

爸爸，生日快樂　　Happy Birthday, Papa

爸爸：

　　你已經離開了 3 個月。還記得在你呼出最後一口氣的時候，我是非常捨不得，但心底裡卻是安慰。因爲你已經不用再每天躺在病牀，無結果的等，看著醫護人員每天在給你做動脈抽血，你痛苦的樣子、無奈的眼神，内心會反問自己「值得嗎」？「還要等幾耐啊」？

　　媽媽也鬆了一口氣，我們也再次聽到她消失了很久的笑聲，媽媽比我們想像中更堅強。還記得在你彌留的時候，我們一家人在病房陪伴著你，醫生說：「是時候了」，護士則叫我們在你耳邊說話，因爲你還是聽見的。對不起，我在你的耳邊真是說不出任何東西，要說的只可以在心裡說著，我只懂得用手輕輕撫摸你燒得燙燙的額頭。

　　被安排得密密麻麻的工作使我 focused，不去想太多。但每當在工作過後，又或是在夜闌人靜的時候，你的樣子、你的聲音就會出現在我的腦海裡。我記得你倆每年的結婚週年紀念日，你會給媽媽買一支玫瑰花，2017 年的聖誕節，我在做你喜歡吃的 cheese toast，你坐在飯枱前慢慢 enjoy 你非常喜歡吃的「邪惡多士」。你不太喜歡說話，但卻非常 attentive 的把我們在飯桌上所講的都聽入耳，大家在說的笑話你也會大笑。你是一個好爸爸、一個好丈夫。你從不會跟我「say no」，永遠都是默默的支持我。Your silence is always a loving one。

　　今天 2 月 17 日，是你的生日。爸爸，祝你生日快樂！好想你。

John Boi 上

Dearest Papa,

It's been three months since you left us. I still remember that day, when the doctor told us "it's about time." We were all there with you. I was touching your forehead, but not brave enough to whisper in your ears the words that I wanted to tell you.

We are all relieved that it's over, and happy that you are no longer suffering. You have been a true fighter and a man of patience. I cannot imagine the feeling of doing nothing, lying in your ward every day, and not able to live freely as you used to. People say life and death is a process, but isn't this a process to test our patience?

I have been keeping myself extremely busy so I have no time to reminisce. But there are days and moments that I will suddenly feel "something is not right" deep down inside my heart. Like the day of your wedding anniversary, remembering how you bought Mama a rose for more than 40 years. And Christmas Day for sure, with all of us coming back home and cooking, and you smiling at us and enjoying the food we cooked. You didn't talk much, but you were listening to all our conversations and laughed at all the jokes.

And of course, it's food that reminds me of the old days. Do you remember the minchi that I made for you? You loved it so much! And the egg custard that I snuck into the hospital, without telling Mama or the nurses. And then there's the cheese toast that you asked for at your last Christmas. I like to remember you sitting comfortably at home, enjoying cheese toast.

Time flies and it's been three months already. The feeling of losing someone dear is truly unbearable, and yet we have to move on and live positively. Today is your birthday, and I just want to tell you: Happy Birthday, Papa.

Love,
John Boi

芝士免治可樂餅
Cheesy Minchi Croquettes

中葡家庭都愛免治肉「Minchi」，我家也不例外。你可看看我第一本書《鹹蝦燦之味》第89頁，當中有關於免治肉的故事和食譜。今次我將免治肉加了些變化，成為了我另一個自創食譜。

Minchi is a dish loved by many Chinese-Portuguese Eurasian families around the world, and my family is no exception. You can read more about the story of minchi on p. 89 of my first book, *The Ham Har Chaan Cookbook*.

材料（10-12人份量）

免治肉

橄欖油	2湯匙
洋蔥	半個，切蓉
免治牛肉	500克
免治豬肉	100克
蒜頭	2瓣，切蓉
豉油	2湯匙
喼汁	1湯匙
糖	2茶匙
鹽和白胡椒粉	適量

可樂餅

焗薯	4個，去皮
莫扎瑞拉芝士	80克，撕細塊
雞蛋	2隻，拂勻
日式麵包糠	8湯匙
米糠油	適量，油炸用

Ingredient (Makes 10 to 12 croquettes)

Minchi

Olive oil	2 tablespoons
White onion, minced	½
Minced beef	500g
Minced pork	100g
Garlic, minced	2 cloves
Soy sauce	2 tablespoons
Worcestershire sauce	1 tablespoon
Sugar	2 teaspoons
Salt and white pepper	to taste

Croquettes

Baking potatoes, peeled	4
Mozzarella cheese, shredded into small pieces	80g
Eggs, beaten	2
Japanese panko	8 tablespoons
Rice bran oil	for deep frying

做法 How to cook

做免治肉：在大的煎鑊中倒入油，以中大火加熱，放入洋蔥炒至軟身透明，加入牛肉、豬肉和蒜頭，炒約6分鐘至肉質金黃。倒入豉油、喼汁和糖，以中火煮約10分鐘至乾身，以鹽和胡椒粉調味，盛起略放涼。

做可樂餅：將焗薯蒸15分鐘至軟身熟透，用壓薯蓉器壓成蓉。

取3湯匙薯蓉壓平於手中，放入2茶匙免治肉及少許莫扎瑞拉芝士，將薯蓉包著芝士肉碎，搓成約2厘米的球狀或橄欖形，共完成12個薯餅，將它們放碟上，放入雪櫃冷藏20至30分鐘。

將蛋液放入淺碟中，麵包糠放另一碟，裹上蛋液，再沾上麵包糠。

在湯鍋中倒入油至約5厘米高，加熱至少許煙冒出，將薯餅逐個放入油中炸2分鐘至金黃，用勺盛起放廚紙上瀝油。

For the minchi: Heat the oil in a large skillet over medium-high heat. Add the onion and sauté until soft and translucent. Add the beef, pork and garlic and cook until the meat is browned, about 6 minutes. Add the soy sauce, Worcestershire sauce, and sugar and cook over medium heat until it becomes dry, about 10 minutes. Season to taste with salt and pepper. Set aside to cool slightly.

For the croquettes: Steam the peeled potatoes until they are soft and cooked, about 15 minutes. Use a potato masher or ricer to mash the potatoes.

Take about 3 tablespoons mashed potato and place about 2 teaspoons minchi and 1 teaspoon shredded mozzarella in the center with your hands, form the potatoes around the minchi and cheese, forming a ball (about 2cm in diameter) or oval around the beef and cheese, like a croquette. Repeat to make about 12 croquettes, placing them on a platter. Cover and refrigerate for 20 to 30 minutes.

Place the beaten egg in a shallow dish and place the panko in a second dish. One by one, dip the croquettes in the egg, and then roll in the panko to cover.

Heat about 5cm oil in a saucepan until you can see light smoke coming out. In batches, add the croquettes to the oil and fry until golden brown, about 2 minutes. Remove with a slotted spoon and drain on paper towels.

芝士多士
Cheese Toast

芝士多士有很多不同的做法，有人說它原創於 1912 年英國蘭卡斯特。香港葡人會所「西洋會所」有超過一百年歷史，它的芝士多士出現在餐牌上數十年，是一款很經典的美食。爸爸每次到西洋會所，總會先叫一份芝士多士，然後慢慢翻餐牌看看吃甚麼。

Some people claim cheese on toast was first created in 1912 in Lancaster, UK. Since then, there have been many ways to prepare cheese toast. The version at Club Lusitano is an all-time-favorite classic dish that's been on the menu for many decades. My papa's regimen was to order cheese toast as soon as he arrived at the club so he could enjoy it while flipping through the menu.

材料（4人份量）		Ingredient (Makes 4 servings)	
半硬芝士3-4款	刨碎	3 or 4 semi-firm cheeses, grated:	
車打芝士	110克	Cheddar	110g
Gouda高達芝士	55克	Gouda	55g
Edam艾登芝士	55克	Edam	55g
卡夫巴馬臣芝士粉	10克	Kraft Parmesan cheese powder	10g
蛋黃	2湯匙，拂勻	Egg yolks, beaten	2 tablespoons
全脂奶	1湯匙	Full-cream milk	1 tablespoon
英式芥末	5茶匙	English mustard	5 teaspoons
喼汁	2茶匙	Worcestershire sauce	2 teaspoons
檸檬汁	1湯匙	Lemon juice	1 teaspoon
糖	2茶匙	Sugar	2 teaspoons
鹽	⅓茶匙	Salt	⅓ teaspoon
方包	4片，切半	Sandwich bread, halved	4 slices

做法 How to cook

將焗爐調至燒烤功能，以180 °C預熱15分鐘。你也可用多士爐，調至高溫。

在大碗中放入刨碎芝士、蛋黃、芥末、喼汁、檸檬汁、糖、鹽，拌勻至成糊狀。將方包去皮切半，將芝士糊抹在每塊麵包上，約3毫米厚。將麵包放焗架上，放入焗爐烤至芝士溶化略焦，約需3至8分鐘。

Turn your oven to broil function and preheat at 180°C for 15 minutes. Alternatively, you can use an oven toaster on high.

In a large mixing bowl, mix the cheeses with the egg yolks, milk, mustard, Worcestershire, lemon juice, sugar, and salt until thick like a puree. Trim the crusts from the bread slices and cut each in half. Spread the cheese mixture onto each slice, about 3mm thick.

Put the bread on a rack and place under the grill for about 3 minutes, until the top turns slightly burnt and cheese mixture is melted. This should take no more than 8 minutes.

早飯 · Breakfast in Japan

早飯 = 早餐。

現代人生活忙碌，早餐吃的都是以簡便爲主，在上班的時段，地鐵站的美心餅店又或是東海堂，總會見到上班一族買一、兩個麵包，再加一盒紙包飲品，這便是他們的「早飯」了。

吃一個 full breakfast 對大部份上班一族來說都是 luxury，要在早餐吃上一碗白飯更是「有點兒過份」。

我喜歡在日本旅遊時吃日式早餐，一碗白飯，再配上熱騰騰的麵豉湯、燒魚和一些漬物，在不用上班，身在外國旅遊的日子吃一頓這樣的早餐感覺很實在、幸福和滿足。

今天高山的天氣還是冷，我們一行人走到這個菜市場，吃了一頓非常簡單的鄉土料理早餐。

這個食堂是家庭式經營，早餐款式和賣相絕對不能跟我每次到東京都喜歡入住的 Andaz Tokyo 和式早餐媲美，但我卻喜歡這裡的現實感，而食堂的一旁正播著電視的早晨直播節目，本地的食客大家聚精會神的看著電視機，然後大聲笑，在廣告時段大家會高談闊論，有些人則繼續吃著他們的早飯，節目又再開始，食堂又再次靜下來，大家又是聚精會神的在看電視。Well，這個情景，此時此地，一切都配合得太美了。

I don't necessarily eat a lot for breakfast, but an exception is when I travel in Japan. I enjoy the Japanese way of breakfast: miso soup, seared or grilled fish fillet, steamed savory egg, plates of little condiments with different pickles, and a bowl of hot rice or porridge.

The feeling of having a bowl of hot rice or porridge is very filling and loving. And I remember how breakfast was called *jo fan* (literally, "rice in the morning") in the good old days. Even now, we like to have porridge in the morning, with a plate of fried noodles on the side.

So I am in this cozy little restaurant in Takayama for breakfast. Yes, it's traditional Japanese breakfast with a plate of quick-seared beef (*gyuniku no tataki*), some pickles, miso soup, and a bowl of white rice. It's nothing fancy but it's so filling and warm, especially after the snow earlier in the morning. The restaurant seems like a family run one, with dining area in the front and living area at the loft.

It's almost 8 o'clock. The television is turned on and showing a local program that I cannot understand. But my eyes are on the screen, along with the other diners, and I laugh when everyone else laughs. Oh yes, I just love this moment.

鹹蝦燦日式家庭早餐——
番茄盅、紅蘿蔔麵豉湯、
地中海烤鯖魚
Japanese Homey Breakfast,
Rocha Style——
Breakfast Tomatoes, Carrot and Miso
Soup and Grilled Mackerel

這是我對傳統日式早餐的演繹，以營養豐富的紅
蘿蔔麵豉湯、炒蛋番茄盅，配以簡單的烤魚，當
然不能欠了一碗白飯或白粥。我加了點西方元
素，你可隨意設計你心目中的版本，或以一杯香
濃咖啡代替綠茶。各位早晨！

This is my own interpretation of a traditional
Japanese breakfast, with a nourishing carrot soup,
tomatoes stuffed with eggs, and simply grilled fish.
Of course, the centerpoint of a Japanese breakfast
is a bowl of hot rice or porridge. I have added
some Western elements—feel free to design your
own Japanese breakfast. Perhaps you'd like to
include a nice cup of strong coffee to replace the
green tea. Good morning and have a nice day!

番茄盅

材料（2人份量）

大圓番茄	2個
雞蛋	2隻
蛋白	2隻
牛奶	70毫升
煙三文魚片	150克，切小塊
鹽和胡椒粉	適量
無鹽牛油	15克
新鮮細香蔥碎	2湯匙

Breakfast Tomatoes

Ingredient (Makes 2 servings)

Large round tomatoes	2
Eggs	2
Egg whites	2
Milk	70ml
Smoked salmon, cut into small chunks	150g
Salt and pepper	to taste
Unsalted butter	15g
Chopped fresh chives	2 tablespoons

做法 How to cook

預熱烤爐。將番茄的頂部橫切去½厘米，棄掉，用湯匙刮去肉和籽，棄掉。將番茄盅放入烤爐內，離熱源5厘米，烤4分鐘至頂部微黃，保溫備用。

在大碗內打入蛋和蛋白，倒入牛奶拌勻，放入三文魚塊和調味拌勻。牛油放入中型易潔鑊內，以小火加熱，倒入蛋液不斷炒至你喜歡的凝固程度（但不要煮得過久，否則會變乾和無味）。

將炒蛋舀放番茄盅內，灑上香蔥碎立刻享用。

Preheat the broiler. Cut off the tops of the tomatoes, about ½cm from the top, and use a tablespoon to scoop out the flesh and seeds. Discard the tops and the flesh and seeds. Place the tomatoes 5cm under the broiler and broil for 4 minutes, until the tops are lightly browned. Set aside and keep warm.

In a large mixing bowl, whisk together the eggs, egg whites, milk, and salmon and season with salt and pepper. Heat the butter in a medium nonstick skillet over low heat. Add the egg mixture and cook, stirring constantly, until the eggs are set to your preference (but do not overcook, otherwise the eggs will be dry and tasteless).

Fill the warm tomatoes with the scrambled egg and salmon mixture, sprinkle the chives over the top, and serve immediately.

紅蘿蔔麵豉湯

材料（2人份量）

		Carrot and Miso Soup	
		Ingredient (Makes 2 servings)	
雞湯	400毫升	Chicken broth	400ml
紅蘿蔔	100克，切片	Carrots, sliced	100g
橄欖油	1湯匙	Olive oil	1 tablespoon
煙肉	20克，切小塊	Bacon, cut into small pieces	20g
西芹	20克，切碎	Celery, chopped	20g
赤麵豉	1茶匙	Red miso paste	1 teaspoon
鹽和胡椒粉	適量	Salt and pepper	to taste
新鮮芫茜碎	適量（裝飾用）	Fresh cilantro, chopped	to garnish

做法 How to cook

雞湯倒入湯鍋中，以中火煮滾，加入紅蘿蔔後蓋好，調至小火煮至紅蘿蔔軟腍熟透。離火，將紅蘿蔔在湯中壓爛。

炒鑊加油，以中大火燒熱，放入煙肉煎至香脆，離火，加入西芹碎拌勻。

將西芹煙肉放入紅蘿蔔湯內煮至滾，離火，加入麵豉拌至溶化，以鹽和胡椒粉調味，灑上芫碎即成。

In a saucepan, bring the chicken broth to a boil over medium heat. Add the sliced carrots and cover. Reduce the heat and simmer until the carrots are soft and cooked. Remove from the heat and mash the cooked carrots in the liquid.

In a frying pan, heat the oil over medium-high heat. Add the bacon and fry until crispy. Remove from the heat and mix in the chopped celery.

Add the bacon and celery to the carrot soup and bring to a boil. Remove from the heat, mix in the miso paste, and stir until the paste is dissolved. Season with salt and pepper, sprinkle with cilantro, and serve.

地中海烤鯖魚   Grilled Mackerel, Mediterranean Style

材料（2人份量）   Ingredient (Makes 2 servings)

檸檬汁	2湯匙	Lemon juice	2 tablespoons
初榨橄欖油	1湯匙	Extra virgin olive oil	1 tablespoon
茴香籽	1茶匙，壓碎	Fennel seeds, crushed	1 teaspoon
辣椒碎	1茶匙	Chili flakes	1 teaspoon
鹽和胡椒粉	適量	Salt and pepper	to taste
鯖魚柳	2塊 （約250-300克一塊）	Mackerel fillets	2 (250 to 300g each)
火箭菜	30克	Rocket leaves	30g
新鮮番茜	20克	Fresh parsley	20g
意大利黑醋	1湯匙	Balsamic vinegar	1 tablespoon

做法 How to cook

碗中放入檸檬汁、橄欖油、茴香籽、辣椒乾、½茶匙鹽和少許胡椒粉拌勻。

將調味料抹在鯖魚柳的兩邊，放室溫醃15分鐘。以中大火加熱平底煎鑊，同時將火箭菜和番茜拌勻後放碟的中間。

將魚柳放煎鑊內，皮向下，煎50秒，反轉再煎40秒至熟，上碟放在火箭菜旁。

將餘下的醃汁與意大利黑醋拌勻，放炒鑊中以中火加熱30秒至微暖。

將醬汁澆在魚和菜上，即時享用。

In a mixing bowl, combine the lemon juice, olive oil, fennel seeds, and chili flakes. Season with ½ teaspoon of salt and a pinch of pepper. Mix together.

Brush some of the marinade on both sides of the mackerel fillets and marinate at room temperature for 15 minutes. Preheat a griddle pan over medium-high heat. Meanwhile, mix the rocket leaves with the parsley and arrange in the center of a plate.

Place the fillets on the hot griddle pan, skin side down, and cook for 50 seconds. Flip the fillets and cook the other side for another 40 seconds, until cooked. Transfer to the plate with rocket leaves.

Combine the remaining marinade and balsamic vinegar in a frying pan and cook over a medium heat for 30 seconds, until it's merely warm.

Spoon the hot dressing over the fish and rocket leaves and serve immediately.

努力的証據 · A Photo

從高山坐了大約 1.5 小時的車到了金澤，在 21 世紀美術館的展覽廳看到一個有關「手」的相片展覽會。這張相片命名爲「努力的証據」，是一位對入廚相當有理想和抱負的廚師的照片。

看著這張相片，百般滋味在心頭，it's not only luck，我是有付出過的。

I know it takes a lot of hard work and dedication, but opportunities count a lot too. This photo from 21st Century Museum of Contemporary Art gives me a lot to think about.

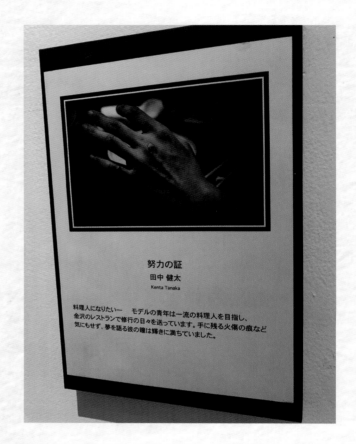

2
0
1
9
·
10/03

溫泉旅館　Onsen Ryokan in Japan

我喜歡日本，尤其是日式溫泉旅館，每年都會專誠到訪日本 2、3 次，住溫泉旅館。

說到溫泉旅館，我更會挑選一些附有私人風呂的入住，因爲我偏愛 exclusivity。今次入住了位於距離高山大約 45 分鐘車程，在福地溫泉區附近的隱庵。隱庵只有 12 間房間，每間房間都有獨立的私人風呂，而整間酒店更是恕不招待 16 歲以下的朋友。

從入住時的介紹，晚上的懷石料理，第二天早上的早餐都是物有所值。最愛的就是浸在自己的私人風呂，周圍還下著微微的細雪。電視上曾經見過的情景終於發生在我眼前上。實在太漂亮了！

對啊！還記得 2016 年 3 月，北海道定山溪溫泉之旅給了我做「鹹蝦燦」的初念。日式私人風呂跟我真是息息相關。

一直在寫著這篇日誌，肚子在「打鼓」。剛才 6 點才吃過的懷石料理好像已經消化了！哎呀，我突然好想吃日式照燒雞啊！爲何大部份頂級的日式溫泉旅館都沒有 room service 提供的？

I love the *onsen ryokan* (hot springs guesthouse) experience and usually make two trips each year to Japan. It's not for shopping, but only for onsen.

I am at the Kakurean Hidaji in Takayama for two nights. It is an exclusive, adults-only onsen with only 12 rooms, each of which comes with its own private outdoor onsen.

 I check in, and it is so quiet and I hardly see any other people. The service is impeccable from the moment I arrive, including a briefing from my hostess about the room, meal requests, and the time for *kaiseki* dinner and breakfast. But the best part is, of course, my private onsen. It is just outside, in my private courtyard, where I can soak myself privately, all by myself. Even better, it's snowing, and the view is just amazing!

But the only thing is that dinner at an onsen ryokan is usually served early. You may have a choice to have dinner as early as 6 p.m. or maybe 7:30. But there isn't much to do after dinner, except onsen.

I have my dinner at 6 p.m. and now it's not even 10 p.m. Oh damn, I feel like having a teriyaki chicken. Why there is no room service?

照燒雞
Teriyaki Chicken

我們很容易在超市買到瓶裝的照燒汁，但我喜歡自己準備，除了配雞，也可配豬、牛、三文魚和鯖魚。

We can easily get a bottle of ready-made teriyaki sauce from supermarket, but I like to make my own. You can also use the homemade sauce for pork, beef, salmon or mackerel.

材料（4人份量）		Ingredient (Makes 4 servings)	
照燒汁		Teriyaki Sauce	
味醂	200毫升	Mirin	200ml
日本豉油	100毫升	Japanese soy sauce	100ml
糖	2湯匙	Sugar	2 tablespoons
雞腿肉		Chicken Thighs	
連皮雞腿肉	500克	Chicken thighs, with skin on	500g
鹽和胡椒粉	適量	Salt and pepper	to taste
橄欖油	1湯匙	Olive oil	1 tablespoon

做法 How to cook

照燒汁：將味醂放小鍋內，以小火加熱2至3分鐘，加入日本豉油和糖，再煮2分鐘，約剩200毫升照燒汁。放雪櫃可保存4天。

雞肉：將雞肉以少許鹽和胡椒粉調味。在煎鑊加油，以中火燒熱，放入雞肉煎至兩面金黃，然後再煎8至10分鐘至雞肉熟透。

加入2至3湯匙照燒汁，煮至有光澤及汁料包裹著雞肉。離火，將雞肉切片，淋上醬汁即成。

For the teriyaki sauce: In a small saucepan, heat the mirin over low heat for 2 to 3 minutes. Add the Japanese soy sauce and sugar and simmer for another 2 minutes. Makes about 200ml sauce. Leftovers can be stored in the refrigerator for up to 4 days.

For the chicken thighs: Lightly season the chicken thighs with salt and pepper. Heat the oil in a frying pan over medium heat. Add the chicken and brown on both sides. Continue to cook, turning, until the chicken is cooked through, 8 to 10 minutes.

Add 2 to 3 tablespoons of the teriyaki sauce and continue to cook until it has glazed and covered the chicken. Remove the pan from the heat. Slice the chicken and serve with any remaining sauce from the pan.

隱庵（福地溫泉區）(Kakurean Hidaji Fukuji Onsen)

@ www.kakurean.com

高山 Best Western (Best Western Takayama)

@ www.bestwestern.co.jp/takayama/

我只有在名古屋住了一晚，在高山 Best Western 則住了五晚，以此為中轉站，前往其他地方比較方便。

I stayed in Nagoya only for a night on my arrival, but stayed in Takayama for a total of 5 nights as my hub to other towns.

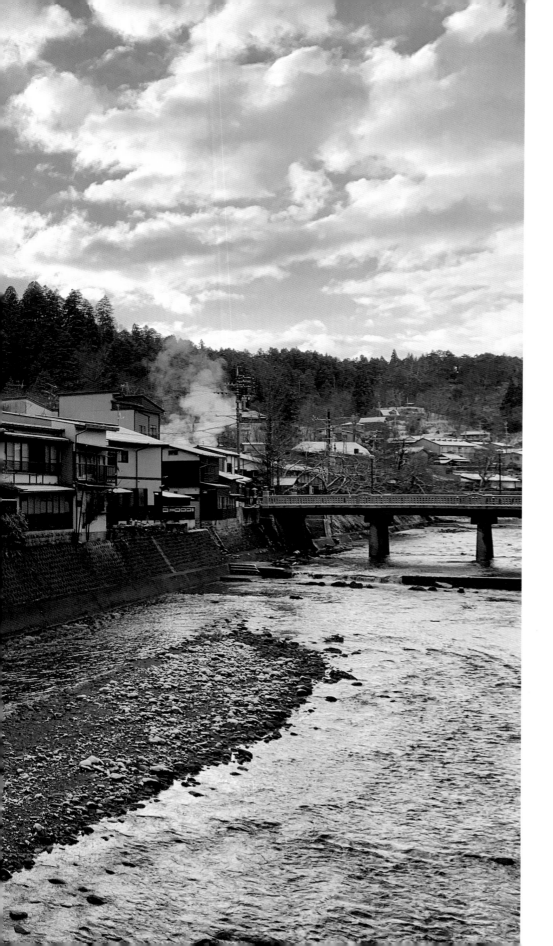

煮食讓我尋回陳小玲　Be Yourself

人生如舞台。在現實生活中，我們扮演多少個角色？在家裡，我是母親的大仔，我是 3 位毛孩的爸爸，我是一位料理導師，我是一位 micro influencer。你呢？李太（Mrs Reis）是我其中一位學生，她的丈夫也是一位「鹹蝦燦」。李太是一名太太、4 名仔女的母親，她更是一位英文會話的補習老師。

上星期在上料理班的時候，有一位素未謀面的同學走到我和李太前面叫著「陳小玲！陳小玲！」我還未給反應的時候，李太已經跟對方點了頭，然後兩人大笑起來。原來那名新同學是李太中學年代的同學，而「陳小玲」就是李太未嫁前的名字啊！瞬息間，他們兩個就「雞啄唔斷」的在 chit chat，久不久就傳來他們的大笑聲。

今天李太來上課，主動跟我說起上星期的事情。

「你知道嗎？這幾年一直忙，爲家庭，爲事業。如果不是 Stella 當天叫了我，我自己也幾乎忘了我的中文名。」

「這個中文名是我在學生年代和最初在上班時才會用，後來嫁了，也跟著用了丈夫的姓氏 Reis。Mrs Reis 已經用了差不多超過 40 年了。」

「我喜歡別人稱我爲陳小玲，因爲陳小玲就是我，而不是一位 grandmother，又或是 mother or wife。上料理班的時候我真的好 enjoy，因爲這 3 小時是我的 me time。」

「John Sir，你以後可以叫我做陳小玲啊！」我聽了李太 Mrs Reis 的話，大家都笑了。

We are all playing many different roles in the real world: a father, a husband, a son, a working man. A mother, a wife, a daughter, a working woman. And I wonder how often we all get our own "me time," and are truly able to be ourselves. As we grow older, we have more and more responsibilities to take on, and it's not likely that we will have only one single role.

Mrs. Reis has been attending my cooking classes for almost a year. She's Chinese with a Portuguese husband, a mother of four, and a grandmother as well. And she works as a English language tutor. She's always busy with making plans for her family and arranging classes for her language students. Last week, we were chatting during the class and suddenly another student called out "Chan Siu Ling!" in front of us. I was a bit uncertain who she was referring to, but Mrs. Reis answered with surprise and joy. It was an old friend from high school, using Mrs. Reis's Chinese name. Throughout the rest of class, they were both so excited, taking photos and chatting about the good old times. So much so that they both didn't really follow my steps for making the potato pancakes and *bife a marrare*. Well, as long as they are happy, it does not matter.

This afternoon, Mrs. Reis, in class as usual, told me how overjoyed she was to meet her old buddy in a cooking class. She told me, "It was so strange to hear someone calling my Chinese name, a name that I haven't heard for a long long time, since I never use it after being known as Mrs. Reis for so many years. It just proves how much I love coming to cooking class where I can be myself, have some me time, and put other matters behind. And please feel free to call me Chan Siu Ling!"

We both laughed.

葡式牛扒伴芥末汁

Bife a Marrare (Portuguese Beef Steak with Mustard Sauce)

這是里斯本咖啡館的傳統名菜。我很愛這菜式，特別是伴炸薯條或薯餅享用，它們跟我介紹的芥末汁很配。Savora 芥末微甜帶辣，是葡萄牙常用的芥末，如買不到，你可以用第戎芥末 Dijon Mustard 代替。

This is a traditional dish served in coffee houses in Lisbon. I love this recipe very much, especially when there are some French fries (the authentic way) or potato pancakes to go with the succulent sauce. Savora is a slightly sweet and spicy French mustard popular in Portugal; Dijon mustard would be a good substitute.

材料（4人份量）		Ingredient (Makes 4 servings)	
無鹽牛油	30克	Unsalted butter	30g
橄欖油	2湯匙	Olive oil	2 tablespoons
肉眼牛扒	4塊，每塊200克	Rib-eye steaks	4 (200g each)
鹽	1茶匙，牛扒調味用	Salt	1 teaspoon, for seasoning the beef
Savora或 第戎芥末醬	2湯匙	Savora or Dijon mustard	2 tablespoons
忌廉	110毫升	Cooking cream	110ml
全脂奶	110毫升	Full-cream milk	110ml
檸檬汁	2湯匙	Lemon juice	2 tablespoons
炸薯條或波蘭薯餅（p.126）	伴食	French fries or Polish Potato Pancakes (p.126)	for serving

做法 How to cook

大煎鑊中放入牛油，以中火加熱至溶。將牛扒以鹽調味，然後放入煎鑊中煎至你喜歡的熟度。我一般每邊煎3分鐘，約半生熟，如每邊煎4分鐘則會6成熟。牛扒煎好後盛起備用。

原鑊放入芥末、忌廉、牛奶和檸檬汁拌勻，以中火煮30秒至混合略稠。

牛扒趁熱伴芥末汁及薯餅享用。

Melt the butter with the olive oil in a large frying pan over medium heat. Season the steaks with salt, add to the pan, and cook until done to your liking. I usually give 3 minutes on each side for rare, or 4 minutes each side for medium. Remove the steaks from the pan and set aside.

In the same frying pan, combine the mustard, cream, milk, and lemon juice and cook over medium heat for 30 seconds, until everything is well combined and thickened slightly.

Serve the steaks hot with the sauce and fries or potato pancakes.

波蘭薯餅
Polish Potato Pancakes

二次世界大戰的時候，波蘭的店舖物資短缺，沒有太多貨品選擇，馬鈴薯餅就成了很重要的食品。時至今日，煎薯餅依然是大人和小朋友的寵兒，配糖及酸忌廉特別美味。煎薯餅也很配葡式牛扒伴芥末汁（p.125）。

Placki Ziemniaczane were especially popular in Poland during World War II when there was very little to buy in the shops. Now, the pan-fried potato cakes are a treat for children and still an all-time-favorite with adults — especially delicious with sugar or sour cream. And they also go very well with Portuguese Beef Steak with Mustard Sauce (p.125).

材料（4人份量）

焗薯	4個，去皮刨碎
洋蔥	1個，切蓉
雞蛋	2隻
麵粉	4湯匙
鹽和白胡椒粉	適量
橄欖油	適量，半煎炸用

Ingredient (Makes 4 servings)

Baking potatoes, peeled and grated	4
White onion, minced	1
Eggs	2
All-purpose flour	4 tablespoons
Salt and white pepper	to taste
Olive oil	for shallow frying

做法 How to cook

在大碗中放入馬鈴薯、洋蔥、雞蛋、麵粉、鹽和胡椒，用手將所有材料拌勻。

炒鑊放入2湯匙橄欖油，以中火加熱。用大湯匙小心舀入一大匙薯仔料，用叉略輕壓扁，煎約5分鐘至底面金黃，然後反轉再煎4分鐘至另一面也金黃，用漏勺將薯餅盛起放廚紙上吸油。共完成約8塊薯餅，可將做好的薯餅放焗爐中保溫。微暖享用。

Combine the potatoes and onion in a large mixing bowl with the eggs, flour, salt, and pepper. It's best to use your hands to combine and mix everything thoroughly.

Heat up a frying pan with about 2 tablespoons olive oil over medium heat. Carefully put a large spoonful of potato mixture into the pan and flatten it slightly with a fork. Fry the pancake until it is golden brown on the bottom, about 5 minutes. Then turn and cook on the other side until browned, about 4 minutes. Remove from the pan with a slotted spoon and drain on paper towels. Repeat to make about 8 pancakes. (You may want to keep the cooked pancakes warm in a warm oven while you preparing the rest of the pancakes.) Serve the pancakes while they are warm.

我的「秘密情人」　My Secret Lover

牛油啊，牛油啊！妳的出現喚起了我對入廚的慾望。是妳太瞭解我的原因，每次做菜妳準會在適當的時候出現，然後將我煮的菜變得錦上添花，汁變得更濃更滑，蛋糕變得輕盈鬆透。

同學們稱妳爲我的「秘密情人」。妳會介意嗎？我知道妳不會的。

牛油啊！妳總是給我帶來了無限的創意，妳實在太善解人意了！教我如何不想妳呢？

後記：OK！我傻了！哈哈哈！不過我對牛油真的是非常偏心，每次做菜都不會缺少的！

I love butter! I use it not just as a spread on toast, but whenever I can when cooking. Frozen butter is a very good way to quickly thicken up the sauce and as a final touch to enhance flavor. Just take a small knob of frozen butter, add to pan juices after cooking meat, and you'll have a rich amazing sauce. As butter is usually used towards the end of cooking, I keep it in the fridge; whenever my students see me taking a knob of butter out, they joke, "Here comes your secret lover!"

I was doing my grocery shopping earlier today and I saw some nice pork chops. I decided to treat myself to the indulgence of pork chops, my favorite Hong Kong – style buttery pasta, and sautéed Chinese kale in butter. It should be so yummy!

港式意大利闊條麵
Hong Kong – Style Fettucine

我喜歡吃意粉，也試過煮不同風味的意粉，例如冬陰忌廉海鮮湯意粉、韓式烤牛肉長通粉及港式意粉。這意粉是香港的經典美食，很多港式西餐廳或茶餐廳都有供應，它的做法其實很容易，而且味道很好，很適合自家製。一般香港餐廳會用意大利粉，但我愛用闊條麵，因為它面積較大可吸收更多汁，也和一般的意粉有所分別。當然不能忘了我的最愛 — 牛油，是這道菜的精髓。

I love pasta. I have tried many fusion pastas and love them all: Tom Yam creamy seafood soup with pasta. Korean BBQ beef with penne. And Hong Kong–style pasta—a classic dish that you can order from many Hong Kong restaurants. But the preparation is so simple, it's easy to make yourself, yet the taste is remarkable. Restaurants in Hong Kong usually use spaghetti, but I like fettucine because the flatter surface blends with the sauce more evenly. Plus, it makes my dish stand out from the usual ones with spaghetti. Hahaha. Of course butter, my secret lover, is the essence to this dish.

材料（2人份量）

意大利闊條麵	200克
橄欖油	1湯匙
洋蔥	1個，切蓉
蒜頭	4瓣，切蓉
茄汁	80克
無鹽牛油	60克
雞湯或水	100毫升
糖	1湯匙
煮熟雞肉腸	100克，切片（可免）
鹽和白胡椒粉	適量

Ingredient (Makes 2 servings)

Fettucine	200g
Olive oil	1 tablespoon
White onion, minced	1
Garlic, minced	4 cloves
Ketchup	80g
Unsalted butter	60g
Chicken broth or water	100ml
Sugar	1 tablespoon
Cooked chicken sausages, sliced	100g (optional)
Salt and white pepper	to taste

做法 How to cook

煮滾一大鍋水，放入鹽及闊條麵，煮至你喜歡的硬度，最好軟硬適中，盛起保溫。

在大炒鑊中以中火加熱橄欖油，放入洋蔥炒至軟身，調至小火，加入蒜頭多煮數十秒，然後加入茄汁、牛油、湯及糖，不斷拌煮5分鐘至汁略稠。

喜歡的話可加入雞肉腸煮2分鐘至熱透，放入闊條麵拌勻至沾滿醬汁，調味即可。

Cook the fettucine in a large pot of salted water to your preference—al dente is preferred. Drain and keep warm. In a large sauté pan, heat the olive oil over medium heat. Add the onion and sauté until softened. Turn the heat to low, add the garlic, and cook for no more than 1 minute. Add the ketchup, butter, broth, and sugar and cook, stirring occasionally, until the sauce becomes thick, about 5 minutes.

If you like, add slices of chicken sausage and cook for another 2 minutes, until heated through. Add the cooked fettucine and toss to make sure the pasta is coated with the sauce. Season to taste.

香煎豬扒
Pan-Fried Pork Chops

這是我其中一個 comfort food 系列，上好的豬扒加少量調味，然後煎至香噴噴，配港式闊條麵（p.131）及牛油炒芥蘭（p.135）。最好用有坑紋的煎鑊，煎時的聲音和香味都令人難忘。

This is one of my favorite comfort foods: a good-quality of pork chops, lightly seasoned and pan-fried, served with Hong Kong–style fettucine (p.131) and sautéed Chinese kale (p.135). Get a nice ridged grill pan if you can and the sizzling sound and smoky flavor are something that you will never forget.

材料（2人份量）		Ingredient (Makes 2 servings)	
走地豬扒	2塊（2厘米厚）每塊150-200克	Free-range pork chops, 2cm thick	2 (150 to 200g each)
海鹽	1湯匙	Sea salt	1 tablespoon
白胡椒粉	1茶匙	Ground white pepper	1 teaspoon
無鹽牛油	2小塊，每塊約20克	Unsalted butter	2 knobs (20g each)
檸檬汁	1湯匙	Lemon juice	1 tablespoon

做法 How to cook

豬扒以鹽和胡椒粉調味，放室溫醃20分鐘。預熱焗爐至180°C。以中火加熱坑紋煎鑊，見冒出煙即放入豬扒，無需放油，將每邊煎約4分鐘至微黃，期間只反轉一次。

用錫紙蓋住煎鑊，將鑊放入焗爐焗5至8分鐘，至豬扒熟透。

將煎鑊取出，在豬扒上放上一小塊牛油及擠上檸檬汁，待2分鐘至牛油溶化，即可享用！

Season the pork chops with salt and pepper and let sit at room temperature for 20 minutes. Preheat the oven to 180°C.

Heat a ridged grill pan over medium heat just until you can see smoke wisping from the pan. Without adding any oil, place the pork chops on the pan and cook, turning once, for about 4 minutes on each side, until lightly browned.

Cover the pan with a lid or aluminum foil, transfer to the oven, and bake for 5 to 8 minutes, until the chops are cooked through.

Take the pan out from oven and top each chop with a knob of butter and some lemon juice. Let rest for 2 minutes, until the butter is melted. Enjoy!

牛油炒芥蘭
Sautéed Chinese Kale in Butter

很多香港人都喜歡吃芥蘭，有些用肉碎蒜蓉炒，有些加薑汁炒，有些灼1-2分鐘後以蠔油伴食。我喜歡配蒜頭，並加了神秘材料令它更特別。

Chinese kale is a delicious vegetable, enjoyed by many in Hong Kong. Some sauté it with meat and minced garlic, others love it with ginger juice, and some simply blanche it a minute or two and serve with oyster sauce. I like to sauté the kale and garlic in my secret lover to give it a different touch.

材料（2人份量）	
橄欖油	1湯匙
芥蘭	400克，切成2至3段
蒜頭	4瓣，切蓉
無鹽牛油	20克
味醂	2湯匙
水	1湯匙
糖	1茶匙
鹽和白胡椒粉	適量
烘香芝麻	1湯匙

Ingredient (Makes 2 servings)	
Olive oil	1 tablespoon
Chinese kale, cut into 2 or 3 sections	400g
Garlic, minced	4 cloves
Unsalted butter	20g
Mirin	2 tablespoons
Water	1 tablespoon
Sugar	1 teaspoon
Salt and white pepper	to taste
Toasted sesame seeds	1 tablespoon

做法 How to cook

炒鑊中以中火加熱橄欖油，放入芥蘭及蒜蓉炒1分鐘。加入牛油、味醂、水和糖，加蓋以中火煮1分鐘，至芥蘭莖底部的微白色褪去。

調味，灑上芝麻即成。

Heat the olive oil in a sauté pan over medium heat. Add the kale and garlic and sauté for 1 minute. Add the butter, mirin, water, and sugar. Cover and cook over medium heat for 1 minute, until the light white color from the bases of the stems is gone.

Season to taste and sprinkle with the toasted sesame seeds.

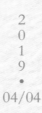

2
0
1
9
·
04/04

<h1>食譜 41　Recipe 41</h1>

在《鹹蝦燦之味》，我在書中「食譜 41」邀請各位分享大家的食譜，而我會揀選其中三位朋友的食譜，放在這本書裡。非常感謝 Tina（p.138）、Serena（p.173）和 Shirley（p.178）的分享，希望各位讀者也喜歡。

In my first cook book, I invited everyone to share their recipes with me. You did, and now I have picked three to feature here in this cookbook. I am so thrilled to present recipes from Tina (p.138), Serena (p.173), and Shirley (p.178). I hope everyone will enjoy reading their recipes and cooking them.

食譜 41
Recipe 41

食譜41是什麼？如果你從頭到尾看過我的書，你應該已經掌握了如何烹調一些葡國菜。現在是時候製作自己的食譜了。邀請你加入我的美食之旅，分享烹飪的樂趣。把你的原創新式葡國食譜分享到我的臉書。我會在第二本烹飪畫中介紹三款由讀者提供的食譜，好好享受烹飪之樂！

What is Reci... if you have re... you probably a... to cook Portugue... it's the time to make ... my Food Journey and ... Post your original ... recipe on my pa... recipes in my ... Enjoy ...

葡式朱古力蛋糕
Portuguese Mud Cake

認識 John Sir 數年，他除了是我的良師，亦成為我的好友。John Sir 對食物之熱誠，能保持傳統葡國家鄉美食的特質，亦常帶出一些自我創作的食品，給我們驚喜。上課時風趣幽默，毫不嚴肅，感覺像和朋友聚會。我慶幸能夠參與他第二本 cookbook 的製作，多謝他的邀請，此份作品也是從 John Sir 傳授的葡國食品中啟發而來，我做了兩款不同造型的朱古力蛋糕，蛋糕的面裝飾是用了翻糖和食用色素來做，希望讀者喜歡。

I have known John for years as both my cooking instructor and personal friend. John is passionate about his cooking and teaching, and you can easily feel his positive vibe when he's around. John inspires everyone by introducing new and improvised dishes from his heritage cooking, and he is always so cheerful and humorous. A big thanks to John for inviting me to share my recipe, a rich mud cake inspired by John's heritage cooking. I have made two versions of the mud cake and decorated each with fondant. I hope everyone will love the mud cake recipe.

材料（4人份量）

		Ingredient (Makes 4 servings)	
無鹽牛油	250克	Unsalted butter	250g
即溶咖啡粉	8茶匙	Instant coffee	8 teaspoons
清水	180毫升	Water	180ml
黑朱古力	250克，切碎	Dark chocolate	250g, chopped
自發粉	150克	Self raising flour	150g
麵粉	150克	All-purpose flour	150g
可可粉	60克	Cocoa powder	60g
食用梳打粉	½茶匙	Bicarbonate soda	½ teaspoon
糖粉	500克	Caster sugar	500g
鹽	1茶匙	Salt	1 teaspoon
雞蛋	4隻，略拂打	Eggs	4, lightly beaten
菜油	8茶匙	Vegetable oil	8 teaspoons
白脫牛奶	125毫升	Buttermilk	125ml

做法 How to cook

預熱焗爐至150°C。

將牛油、咖啡粉和水放入湯鍋內，以中火加熱至溶化，離火，加入朱古力碎拌勻，略放涼。

在大碗內篩入所有自發粉、麵粉、可可粉和梳打粉，加入糖、鹽和雞蛋，用打蛋器慢慢打勻，倒入朱古力混合物、油和白脱牛奶，繼續攪拌約2分鐘，或至所有材料完全混和。

將麵糊倒入22厘米（9吋）圓形蛋糕盆，或20厘米（8吋）正方形蛋糕盆內，放入焗爐焗最少55分鐘，用竹簽插入蛋糕試熟。

Preheat the oven to 150°C.

In medium saucepan, cook the butter, instant coffee, and water over medium heat until the butter has melted. Remove from the heat and add the chocolate, mixing well; leave to slightly cool.

In a large mixing bowl, sift together the flours, cocoa, and bicarbonate of soda. Stir in the sugar, salt, and eggs. Whisking slowly, pour in the chocolate mixture, oil, and buttermilk. Mix well for 2 minutes, or until well combined.

Pour the batter into a 22cm (9-inch) round baking pan, or 20cm (8-inch) square baking pan. Bake until a toothpick inserted into the center comes out clean, about 55 minutes.

食譜由 Tina Chung 設計
Recipe by Tina Chung

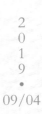

我的晚餐　My Dinner

同學們今天晚上做完了葡式烤排骨（p.231），大家懷著興奮的心情將菜餚帶回家分享。在上堂會說會笑的我，下課後總是變得非常沉默，因為實在有一點累啊！

雖然是有點累，但我卻非常 enjoy 下課後回家路上的感覺，想一想剛才上課的情況作自我檢討，聽住自己喜歡的歌曲，回覆 whatsapp。這種晚間放工回家的景象來得很浪漫呢！

回家太晚，我都不想再做菜了。很多時都會吃個即食麵便是了。同學們知道了說痛心，還嚷著我應該叫工人姐姐為我準備美味的晚餐！

哈哈，大家不用擔心，即食麵也可以煮得「好好味」，而我也當然會將即食麵略加「小小 idea」，變得不平凡。

就如我經常掛在口邊的說話：「入廚可以很簡單，隨心的煮，不用想太多！」

Class finished at 9:30 p.m.

I feel knackered after my cooking class. I am talkative and super hyper during class with jokes and stories. But after class, I am mute. Dead silent. LOL.

The trip home is always an enjoyable one, listening to my favorite music from Spotify, posting photos onto social media, and of course answering all the "unanswered" messages from WhatsApp. I wonder what other people do during their commute home after a long day?

We made Portuguese ribs (p.231) in class tonight, always one of my favorite dishes to teach. My students love packing up the ribs and taking them home for dinner. As for me, I seldom eat the dishes that I taught after class… I don't know… I just want something different.

But what should I have for dinner tonight? Well, by the time I get home, it will be almost 11 p.m.! Should I cook something? Something easy, like pasta? Okay, how about instant noodles. Hahaha… a lot of fans and students made certain remarks about my love of instant noodles. Okay, okay, let me share with you the way I make instant noodles. Maybe you will love them too!

即食麵炒蛋
Instant Noodles in Scrambled Eggs

我愛即食麵！快捷、方便而且 comforting，特別是當你忽然想吃一碗熱湯麵的時候。我用即食麵創作了新煮法，大家也試試看做來嚐一嚐。

I love instant noodles! They are quick, easy, and very comforting, especially when you have a sudden craving for a bowl of hot noodles. I give instant noodles a new twist, similar to the way Japanese yakisoba is prepared. . Proof that cooking can be so much fun. See for yourself!

材料（2人份量）		Ingredient (Makes 2 servings)	
雞蛋	4隻	Eggs	4
牛奶	60毫升	Milk	60ml
忌廉	20毫升	Cooking cream	20ml
鹽和胡椒粉	適量	Salt and pepper	to taste
無印良品即食拉麵	60克（2包）	Muji instant ramen noodle	60g (2 packets)
橄欖油	1湯匙	Olive oil	1 tablespoon
茄汁	適量	Ketchup	for serving
新鮮番茜碎	適量	Fresh parsley, chopped	for serving
木魚乾	適量	Bonito flakes	for serving

做法 How to cook

將雞蛋放大碗內拂勻，加入牛奶、忌廉、少許鹽和胡椒粉拌勻。

用手將即食麵捏成碎塊，放入蛋液中拌勻。

以中火加熱易潔煎鑊，然後調至小火，加入橄欖油，倒入即食麵蛋漿，用木匙不斷從鑊邊推向鑊中心，約炒1分鐘。離火，繼續推炒至炒蛋成形但仍嫩滑。

上碟，加上茄汁、番茜碎和木魚乾即可。

In a mixing bowl, whisk the eggs well. Add the milk, cream, and a pinch of salt and pepper.

Use your hand to break the instant noodles into chunks, then mix into the egg mixture.

Heat a nonstick frying pan over medium heat. When the pan is hot enough, turn the heat to low and add the olive oil. Pour the beaten eggs and noodles into the pan and cook, stirring with a wooden spoon and moving the eggs and noodles from the edges of the pan toward the center, for 1 minute. Remove the pan from the heat and continue to stir until the eggs are scrambled but still look runny.

Transfer to plates. Finish off with ketchup, chopped parsley, and bonito flakes.

我愛用無印良品的即食拉麵，
也可用福字牌的上湯拉麵，
它 90 克一包，取 2/3，加 4 隻雞蛋。

I love the instant noodles from Muji. You may like to replace them with Fuku superior soup instant noodles. However, they come in packets of 90g, so you will need to break off two-thirds to get the amount needed for a serving of 4 eggs.

繼續煮

Cooking Is Never Ending

再次踏足葡國這片土地
百感交集。
活在這大時代，
頓覺無奈、無助。
不，我要活出快樂人生。
煮
給我放下來、靜思、領悟。
感恩、知足常樂。

Doing what you like is freedom,
Liking what you do is happiness.

重新愛上了 Fado　Fado in Lisbon

籌備了超過半年時間的葡萄牙美食之旅開始了！

剛 check-in 了 Atlis Belem Hotel & Spa。大家已經急不及待的去打卡、拍照。看見大家這般雀躍，我都可以放下心頭大石。

今天晚上安排了帶各位到位於里斯本一間非常馳名的餐廳 Adega Machado，這裡的食物不是其吸引之處，其亮點就是富有特色的 Fado 演唱。

中國有戲曲，意大利有馳名歌劇，而葡萄牙則有富有民族特色的 Fado 音樂。記得小時候，嫲嫲總是喜歡在下午放著 Amalia Rodrigues 的 Fado 音樂，喝著一小杯砵酒，閱讀她喜歡的 Agatha Christie 偵探小說。每次我走到她的身旁，她總會告訴我歌詞的意思，Fado 的歌詞大部份都是有關民間疾苦、失戀的 sad stories。說句老實話，當年我覺得這都是較爲「老土」的音樂。

自嫲嫲離開後，慢慢再聽回 Fado，在開始對葡國菜有更深入認識的時候，我再次愛上 Fado。在我的 Spotify 音樂的 playlist 中，有一個專門錄了 Fado 音樂家的清單。

所以我決定在這個葡國之旅的首天就帶各位感受 Fado 的魅力。

在進餐途中，餐廳內的燈徐徐暗下來，也意味著表演快開始了。表演者是一名男性歌手，他開始唱出一些似曾相識的旋律，同學們非常的 enjoy，望望四周，just cannot believe 我以老師的身份帶了同學到葡萄牙。

歌聲是凄怨澎湃，雖然我聽不懂歌詞，但這一刻我是被感動的，我的心底在說：「同學們，多謝各位跟我一起來到里斯本！」

Although I remember watching my avo (grandma) reading her favorite Agatha Christie detective novel, sipping a glass of vintage Porto, and listening to the Fado music of the legendary Amália Rodrigues, I didn't truly know how to appreciate Fado music until I started my culinary career in 2017. Now I want to know about Portuguese cooking, more about the culture, the language, the music, the people, the history.

Fado, a traditional genre of Portuguese music that originated in Lisbon in the 19th century, is known for expressiveness, as well as being profoundly melancholic. Avo would tell me stories of the songs, and how hard life was for Portuguese in those days. She just loved spending her afternoons listening to the music and telling me stories.

I want to impress everyone by organizing a fado dinner tonight. Adega Machado is a very popular place for food and fado, and everyone is looking forward to the evening despite our jet lag. As for the dinner, the caldo verde could have been better with less salt, but the roasted octopus was tender and cooked just perfect. After our dinner, the lights dimmed and then came the moment I have been waiting for. Fado singers come to the center of the restaurant and start signing. It is beautiful and sad. Can't believe we are in Lisbon!

烤八爪魚
Roasted Octopus

在里斯本的第一晚，我們在 Adega Machado 吃了烤八爪魚，深得大家喜愛。我喜歡伴薯仔一同享用，因為那湯汁很美味。配一杯冰凍白酒就最完美。

We had roasted octopus on the first night in Lisbon at Adega Machado. It is such a pleasant dish that is welcomed by everyone. I love to serve it with extra potatoes because the sauce is so yummy. A nice glass of chilled white wine alongside is just perfect.

材料（2人份量）		Ingredient (Makes 2 servings)	
橄欖油	150毫升	Olive oil	150ml
洋蔥	200克，切片	Onions, sliced	200g
連枝番茄	100克，切片	Vine tomatoes, sliced	100g
香葉	2片	Bay leaves	2
蒜頭	6瓣，切片	Garlic, sliced	6 cloves
中型八爪魚鬚	500克，洗淨	Medium octopus legs, rinsed	500g
新薯	350克	New potatoes	350g
法國香草束（香葉、百里香、迷迭香）	1束	Bouquet garni (bay leaf, thyme, and rosemary)	1
鹽和胡椒粉	適量	Salt and pepper	to taste

做法 How to cook

預熱焗爐至180°C。在大焗爐盤內澆上大量的橄欖油，約125毫升，然後放入洋蔥、番茄、香葉和蒜頭，拌至混合。將八爪魚鬚放在上面，再澆上10毫升橄欖油。

用錫紙包好焗盤，放入焗爐焗2小時，至八爪魚軟腍。

約焗1小時15分鐘後，將新薯和餘下的15毫升橄欖油拌勻，放入另一個焗盤內，以鹽和胡椒粉調味，放入香草束。將薯仔放入焗爐焗40分鐘至軟腍熟透。

薯仔與八爪魚一同享用。

Preheat the oven to 180°C. On a large baking sheet, pour a generous quantity of olive oil (125ml), then add the onions, tomatoes, bay leaves, and garlic. Toss around to ensure everything is mixed thoroughly. Lay the octopus legs over the ingredients and drizzle 10ml olive oil on the top.

Cover with foil and roast for 2 hours, until the octopus is tender.

After about 1 hour 15 minutes, mix the potatoes with the remaining 15ml olive oil on a separate baking sheet. Add the bouquet garni and season the potatoes with salt and pepper. Roast the potatoes in the hot oven for about 40 minutes, until the potatoes are cooked and soft.

Serve the potatoes with the octopus legs.

好 Chillax 的一頓午餐　　A Long Relaxing Lunch in Porto

我是一名 control freak，凡事都喜歡親力親為，所以葡萄牙美食之旅真的是做到透不過氣，酒店的電視一直沒有開過，幸好助手 Timothy 也有隨行，他絕對是個好幫手！

說到底，我舉辦的稱為 elite travel，價錢比市面上的貴，所以同學的 expectation 提高絕對是合情合理。而我自己在旅遊和入廚又是一個較為 picky 的人，所以就是忙、忙、忙。

今天是星期日，也是葡萄牙的母親節，我們一行人首先到了 Porto 馳名打卡點 Mosteiro Da Serra de Pilar 拍照，然後便去到 Restaurant Os Lusiadas 吃海鮮午餐。被疊起三層高的海鮮拼盤實在是太吸引了，龍蝦、蜆、蝦、海膽，還有不同的螺類，而那條海鹽焗鱸魚更是一絕。

原本 1.5 小時的午餐，我們共吃了 3 小時，哈哈哈！今天過得很 chillax，還記得同學們今早一起在海邊打卡，然後唱起張惠妹的《聽海》。你們每一位實在太可愛了！

I have been so busy since arriving in Portugal on May 1 that I haven't had the time to turn on the television in my hotel room. I love to watch local TV commercials when I travel as this is one way to learn different cultures.

We had such a nice time together this morning at Mosteiro da Serra de Pilar. Countless selfies, Facebook Live and group photos… so glad we all love taking photos.

The lunch at Restaurante Os Lusiadas was just perfect. We ordered only two dishes: the seafood platter and the very famous baked sea bass in salt crust. The seafood platter is stacked up almost 2 feet tall with lobster, crab, shrimp, prawns, barnacles, whelks, clams, see urchins, and oysters. And when the baked sea bass is served, everyone had their phones ready to capture the moment of the cracking of salted crust. This is not a difficult dish to make, but it's certainly a good way to impress your guests!

Our lunch lasted three hours and it was almost 3 p.m. when we leave the restaurant. Oh, it's Mothers' Day today in Portugal. Happy Mothers' Day Mama!

鹽焗海鱸魚
Salt Baked Sea Bass

這道菜式很難掌握嗎？ 絕對不會，你只需在魚檔購買一條新鮮的魚和廚房備有一個焗爐便行了。這個菜式絕對會為家人和朋友帶來驚喜。

Do I find it difficult to handle this dish? Not really. All you need is to make sure you get a nice and fresh fish from the fishmonger and an oven at home. I love to make this dish to wow everyone.

材料（4-6人份量）		Ingredient (Makes 4 to 6 servings)	
海鱸魚	1條，約1 – 1.8公斤	Sea bass	1 to 1.8kg
新鮮百里香	一把	Fresh thyme	a handful
新鮮迷迭香	一把	Fresh rosemary	a handful
新鮮番茜	一把	Fresh parsley	a handful
海鹽	1.5公斤	Sea salt	1.5kg
幼鹽	500克	Fine salt	500g
蛋白	2隻	Egg whites	2

做法 How to cook

預熱焗爐至180°C。鱸魚保留魚鱗，洗淨，特別是魚腔，然後塞入新鮮香草。

在大碗中倒入海鹽和幼鹽，加入蛋白拌勻成糊狀。將魚放焗盤上，用蛋白鹽蓋滿整條魚，不要讓空氣入內。

將魚放入焗爐焗20分鐘。取出，輕手敲碎鹽殼，小心起出魚皮棄去，取出魚肉，淋上特級初榨橄欖油即可享用。可配焗薯或蔬菜沙律一起享用。

Preheat the oven to 180°C. Clean the fish without scaling it and wash the inside thoroughly. Place the fresh herbs inside the fish cavity.

In a large mixing bowl, mix the sea salt, fine salt, and egg whites together until it becomes a paste. Put the sea bass on a baking sheet and cover well with the salt paste to prevent air from getting in.

Roast in the oven for about 20 minutes. Remove from the oven and tap salt crust gently to remove without damaging the fish. Carefully remove the fish fillet from the bone. Do not serve the fish skin as it's very salty. Drizzle with some good-quality extra virgin olive oil. You may like to have some baked potatoes or green salad on the side. Enjoy!

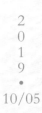
旅程尾聲　Coming to An End

　　凌晨 2 時！我才剛剛把行李執好，明天便要回港了，今天晚上助手預先提前向我請假，已經飛到西班牙繼續 holiday。望住這已經住了 6 晚的房間，剎那間有點兒傷感，是明天要離開葡萄牙的緣故吧。

　　昨天導遊帶我們嚐了葡式點心 petisco，食物做得非常精彩，同學們也玩得非常盡興，整晚就是充滿了我們這班來自香港的大笑聲。

　　能夠跟同學一起出走香港，來到非常遙遠的葡萄牙一起生活，一起上 cooking class，這絕對是一個非常難忘的經驗，盼望 2020 年的 5 月，我再帶同學們到葡萄牙，跟我一起吃喝玩樂。

Shop till you drop! Oh yes, I have just finished packing my bags and I have bought so much in these ten days: olive oil, wine, little stuff for my new home, the very famous soaps and canned foods… aprons… bags… and books!

Last night, Pedro, our tour guide for the trip, took us to a very cozy restaurant 40 minutes away from Porto for *petiscos*, a Portuguese version of tapas. *Petisco* is a snack, something to nibble while having a beer or a glass of wine with friends. The food was amazing and everyone enjoyed it so much. I lost count how many bottles of wine we had and the choice of *petiscos* was endless.

The trip is absolutely exhausting, but I feel so happy seeing my cooking fans loving it. And OMG, I have over 2,000 photos taken just on this trip. Can my publisher be so generous to give me extra pages to print these photos on my book? LOL.

It's almost 2 a.m. here in Porto. Time to sleep!

沙甸魚牛油醬
Sardine Butter

在我第一本書《鹹蝦燦之味》中，介紹了一道美味的砵酒吞拿魚醬（p.203），粉絲們都很欣賞。今次介紹另一款同樣非常邪惡的醬，也是葡萄牙的經典小食之一，希望深受歡迎。

In my first book *The Ham Har Chaan Cookbook*, I introduced the delicious Tuna Spread (p.203) and fans loved it. In this book, I want to introduce another spread that I am sure all will love as well. And it's one of the all-time-favorite petiscos in Portugal.

材料（10人份量）	
無鹽牛油	140克
罐頭沙甸魚柳（橄欖油浸最好）	125克
檸檬汁	2湯匙
新鮮細蔥碎	2湯匙
鹽和胡椒粉	適量

Ingredient (Makes 10 servings)	
Unsalted butter	140g
Canned sardine fillets (preferably in olive oil)	125g
Lemon juice	2 tablespoons
Fresh chives, minced	2 tablespoons
Salt and pepper	to taste

做法 How to cook

將牛油放碗中，將沙甸魚瀝乾油後放入牛油內，用叉壓爛至混合，加入檸檬汁繼續拌勻。灑上細蔥碎，以鹽和胡椒粉調味。

可伴麵包或餅乾享用，或直接用匙舀入口中，妙絕！

Place butter in a medium bowl. Drain sardines, add to the butter, and use a fork to mash together. Add the lemon juice and keep mashing until thoroughly incorporated. Add the chives and season to taste with salt and pepper.

Serve with bread or crackers, or simply scoop with a teaspoon and put in your mouth. LOL.

烤羊肉丸串伴薄荷乳酪
Skewered Meatballs with Minted Yogurt

這是另一道葡萄牙小食，做法容易。有時我會預備更多薄荷乳酪來配墨西哥粟米片。

This is another petisco that I love to serve for entertaining as it is so easy to make. Sometimes I'll make an extra bowl of minted yogurt to go with nacho chips.

材料（4人份量）

		Ingredient (Makes 4 servings)	
免治瘦羊肉	320克	Lean lamb, minced	320g
洋蔥	1個，切碎	Onion, chopped	1
蒜頭	2瓣，切碎	Garlic, chopped	2 cloves
乾紅椒碎	1茶匙	Dried chili flakes	1 teaspoon
新鮮薄荷葉碎	2湯匙	Chopped fresh mint	2 tablespoons
鹽	適量	Salt	to taste
車厘茄	12粒	Cherry tomatoes	12
橄欖油	4湯匙	Olive oil	4 tablespoons

薄荷乳酪

		Minted Yogurt	
低脂原味乳酪	200毫升	Plain low-fat yogurt	200ml
新鮮薄荷葉碎	3湯匙	Chopped fresh mint	3 tablespoons
乾牛至	1茶匙	Dried oregano	1 teaspoon
鹽和胡椒粉	適量	Salt and pepper	to taste

做法 How to cook

以180°C預熱燒烤爐15分鐘。在大碗中放入羊肉、洋蔥、蒜、辣椒碎、薄荷碎拌勻，加鹽，用手將所有材料拌至混合。將肉碎搓成肉丸。

用燒烤串先將羊肉丸串起，再串上車厘茄，然後梅花間竹地串上肉丸及車厘茄（請根據燒烤串的長度調整）。

將肉丸放在焗盤上，掃上油，烤10分鐘，烤至金黃熟透。

烤肉的同時將乳酪、薄荷、牛至拌勻，再加鹽和胡椒粉調味。

Preheat the broiler to 180°C for 15 minutes. Combine the lamb in a large mixing bowl with the onion, garlic, chili flakes, and mint. Season with salt and use your hand to mix everything together.

Thread meatball onto a skewer, followed by a cherry tomato and continue until you have meatballs and tomatoes on each skewer (amount will depend on the length of your skewers).

Transfer the skewers to a baking sheet and brush with the oil. Place under the broiler for about 10 minutes until browned all over and cooked through.

Pour the yogurt into a bowl and add the mint and oregano. Season with salt and pepper and mix together.

里斯本（Lisbon）Atlis Belem Hotel & Spa

📞 (+351) 21 040 0200　　@ www.altishotels.com

花地瑪（Fatima）Hotel Santa Maria

📞 (+351) 24 953 0110　　@ www.hotelsmaria.com

波圖（Porto）Maison Albar Hotels Le Monumental Palace

📞 (+351) 22 766 24100　　@ maison-albar-hotels-le-monumental-palace.com

里斯本（Lisbon）Adega Machado

@ www.adegamachado.pt/en

里斯本（Lisbon）Palacio Chiado

@ www.palaciochiado.pt

花地瑪（Fatima）Laterna do Fado

@ www.laternadofado.makro.rest/

波圖（Porto）Restaurante Os Lusiadas

@ www.restaurantelusiadas.com

波圖（Porto）Conga Casa Das Bifanas

@ www.conga.pt/

心想事成　Dreams Come True

甚麼叫做心想事成呢？我現在真的親身體會到啦。

葡萄牙是我心目中理想旅遊的其中一個國家，心中正在盤算著和甚麼朋友一起去玩呢？參加旅行團行程又趕，團體餐又不好吃，怎算呢？最好旅遊途中可以加插一些特別的事情可以做，而且又好玩的，居然……竟然……突然之間，跑出一位青年才俊，他又識煮，又識嘆，又識食，又好玩得、要求又高、又有架子。他就是現時城中熱辣辣的烹飪導師 John Sir，我們只要報名參加他精心策劃的旅行團，就可以安心、開心地去玩啦！

因為之前曾經跟隨 John Sir 去過馬來西亞及泰國，無論行程與導遊，食物方面，我自己個人都感到非常滿意，這次去葡萄牙，一定又有驚喜喔。

我最欣賞人家用心去做每一件事，這趟旅程讓我見識到葡萄牙人做事的態度熱誠及對朋友的那份熱情。第一次在市場內（Time Out Market）學習烹煮食物，一邊飲著比開水還要平宜的白酒，一邊聆聽老師的指導，一邊吃著老師自家製的麵包，一邊自己在製作，煮完之後，就要自食其果啦，所以要界心機喔，多麼的美好回憶時光。

雖然我個人不是太喜歡食葡撻，但是當我來到 Nata Pura 葡萄牙的總部，去跟葡撻大師學習一些葡撻焗製及技巧，原來製造葡撻並不簡單，可以有很多的變化及款式，這個課堂令我有不枉此行的感覺。回家後一定要繼續練習啊！

最後我們在 El Al Townhouse 學習新派的葡式菜式，由餐前飲料、湯及主菜，甜品等我們都一一學到，用自家種嘅果菜、香料等去烹調食物，又新鮮又環保又好吃。

可以和一班趣味相投的朋友一同去旅遊，一同去學習烹調，何等的賞心樂事，一生難忘。

由同學 Serena Lo 撰文

I always wanted to travel to Portugal, but I also wanted to find some friends to go with me. Then I saw John posted a message about his Gastronomy Tour in Portugal. Without thinking too much, I signed up for it!

I had taken John's similar gastronomy tour in Malaysia and Thailand before, so I knew the tour would definitely meet my expectations, and beyond. The cooking class at Time Out was so amazing. The cooking studio is located within an enclosed market with people watching us throughout the session. Sipping a nice glass of chilled white wine and cooking the dish that instructor taught us was simply relaxing.

Although I am not a big fan of Portuguese Egg Tart, the workshop at Nata Pura was exceptional. Chef Francisco Rocha taught exclusively for our group and we learned some little tricks on how to make good egg tarts.

The cooking class at El Al Townhouse was different. The class covered modern Portuguese cooking and we got to pick our herbs and produce from their herb garden, and then prepared our own dinner.

Traveling with a group of friends and joining cooking classes while on the road is really an unforgettable experience. I shall always remember this amazing trip to Portugal.

By Serena Lo

葡姐三文魚 Macanese Salmon

當我看完 John Sir《鹹蝦燦之味》食譜 41 的文章後，我立即跟他聯絡，希望可以分享這個我很多年前從一名葡籍朋友學的三文魚食譜。當初我還不太清楚這道菜式究竟是葡國菜或澳葡菜，當 John Sir 看過食譜便跟我説這是百分百澳葡菜，因為葡國菜是不會用上生抽和老抽的。這個食譜帶給我很多美麗的回憶，也希望各位會喜歡這菜式。

I got this recipe many years ago from a Portuguese friend. As soon as I saw in John's first book that he was looking for fans to contribute recipes for his second book, I contacted him immediately. When he read my recipe, he told me this is definitely a Macanese dish because of the use of soy and dark soy sauce. I hope you will love this dish! Happy cooking!

材料（4人份量）		Ingredient (Makes 4 servings)	
三文魚柳	450g	Salmon fillet	450g
洋蔥，切片	1個	Onion, sliced	1
番茄，切片	2個	Tomatoes, sliced	2
蒜蓉	10粒	Garlic, minced	10 cloves
糖	2茶匙	Sugar	2 teaspoons
生抽	2湯匙	Soy sauce	2 tablespoons
老抽	1湯匙	Dark soy sauce	1 tablespoon
黑胡椒	4湯匙	Ground black pepper	4 tablespoons
水	1-1½杯	Water	1-1½ cups

做法 How to cook

三文魚兩邊用中火略煎，每邊大約煎2分鐘。放洋蔥略炒至軟身，然後加入番茄和蒜蓉，略炒1分鐘。

放入糖、生抽、老柚、黑胡椒及水，邊煮邊在魚面上淋汁，至收乾汁即成 。

In a skillet, cook the salmon over medium heat, turning once, until lightly browned, about 4 minutes. Add the onion and sauté until soft, about 5 minutes. Add the tomatoes and minced garlic and sauté for 1 minute.

Add the sugar, soy sauce, dark soy sauce, black pepper, and water. Continue to cook, spooning the liquid over the salmon, until the sauce has reduced by half and the salmon is cooked through.

食譜由 Serena Lo 設計
Recipe by Serena Lo

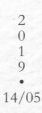

美味食物　It's Delicious !

　　John 曾經給我解釋罐頭食品在西方國家的用法和其文化。在葡萄牙里斯本，他帶我們到訪一餐廳，所有菜式都是由罐頭食物做出來，它的沙甸魚十分好吃！今次是我第一次跟人分享食譜，很緊張。我喜歡快捷簡單的烹飪菜式，就和 John 的入廚理念一樣。希望你會喜歡我這兩道葡式小食。

In Portugal, John explained the use of canned food in Western cooking and took us to a restaurant in Lisbon where all the dishes used canned food. The sardines were great! This is my first time sharing my recipes so I feel so excited. My cooking philosophy is quick and easy, which is exactly the same as John's. I hope you will enjoy my interpretation of Portuguese tapas after my trip to Portugal.

沙甸魚多士 Sardines on Toast

材料（4人份量）

罐頭沙甸魚	125克
忌廉芝士	6湯匙
蒔蘿	2茶匙
橄欖油	1湯匙
鹽和胡椒粉	適量
薄脆多士	8片
車厘茄	4粒，切半

Ingredient (Makes 4 servings)

Sardine fish (in can)	125g
Cream Cheese	6 tablespoons
Dill	2 teaspoons
Olive oil	1 tablespoon
Salt and pepper	to taste
Melba Toast	8
Cherry tomatoes	4, halved

做法 How to cook

將沙甸魚取出瀝乾油。

在碗中放入忌廉芝士、蒔蘿和橄欖油拌成幼滑的忌廉狀，加調味。

將多士薄薄抹上一層忌廉芝士，放上一塊沙甸魚，以車厘茄和額外的蒔蘿裝飾即成。

Remove the sardines from the can and discard the oil.

In a mixing bowl, mix the cream cheese, dill, and olive oil until smooth and creamy. Season to taste with salt and pepper.

Spread a thin layer of the cream cheese mixture on the Melba toasts, followed by a piece of sardine. Garnish with cherry tomato and a bit more dill.

食譜由 Shirley Cheung 設計
Recipe by Shirley Cheung

蟹肉芒果沙律 Crabmeat Mango Salad

材料（4人份量）		Ingredient (Makes 4 servings)	
紅洋蔥，切粒	半個	Red onion, diced	½
西芹，切粒	50克	Celery, diced	50g
蟹肉	300克	Crabmeat	300g
蛋黃醬	¾杯	Mayonnaise	¾ cup
茄汁	2湯匙	Ketchup	2 tablespoons
甜紅椒粉	3茶匙	Paprika	3 teaspoons
英式芥末	1湯匙	English Mustard	1 tablespoon
鹽和胡椒粉	適量	Salt and pepper	to taste
芒果，切粒	50克	Mango, diced	50g
菊苣	10片	Endive	10
蟹子	3湯匙	Crab roe	3 tablespoons
番茜碎	1湯匙	Parsley, chopped	1 tablespoon
Tabasco辣椒仔辣汁	1湯匙	Tabasco	1 tablespoon

做法 How to cook

大碗內放入紅洋蔥、西芹、蟹肉、蛋黃醬、茄汁、甜紅椒粉和英式芥末徹底拌勻，加調味。加入芒果粒拌勻，放雪櫃冷藏1小時。

將一湯匙蟹肉沙律放在一片菊苣上，以蟹子和番茜裝飾。

愛吃辣的話可加數滴Tabasco辣椒仔辣汁。

In a large mixing bowl, combine the onion, celery, crabmeat, mayonnaise, ketchup, paprika, and mustard until mixed thoroughly. Season to taste with salt and pepper. Mix the mango into the crab mixture. Refrigerate for 1 hour.

Scoop the crab mixture onto the endive pieces and then top each with crab roe and parsley for garnishing.

Add a few drops of Tabasco if you love it spicy.

食譜由 Shirley Cheung 設計
Recipe by Shirley Cheung

永平寺　Eihei-ji in Japan

上次在首爾旅遊的時候曾經看過有關 temple stay 的資料，之後在網上也有閱讀過有關在日本禪修的文章。當朋友問我會否有興趣參加位於日本加賀溫泉區附近永平寺的禪修活動，我立刻答應了。

這也是我生平第一次在廟宇渡一宵。由於上兩星期還在葡萄牙，對今次的禪修活動沒有做太多的資料搜集，我只知道入住的不是甚麼酒店，吃的是素菜，心態當然跟上星期在五星級酒店待遇有調整啊！

永平寺雖然是 super huge，但我們不可以隨處走動，到甚麼的地方都要預先通知負責招待我們的僧侶。

晚餐在 5 時半，在晚飯前我們就會先進行 30 分鐘的坐禪活動，老師跟我說不用太刻意的去不想任何東西，只要隨心的便行了，在禪修的 30 分鐘，我滿腦子都是前幾天在葡萄牙的片段。

我們按次序的排隊進入飯廳，從遠處我已經看見一個又一個盛了素菜的盤子放在枱上。嘩！I need to take a picture，我心在想，坐下來，我們一起頌經，然後大家開始進食，盤裡盛著非常精緻的食物，大部份都是用豆腐和南瓜來做的菜式。而其中一個用豆腐做的菜式也啟發了我做一個用芝士和日式豆腐做的 fusion 菜。

但原來永平寺其中一個 house rule 是進餐時不可以交談，而進食的速度不可以太快，也不可以太慢。我看看四周，大家都變成「低頭一族」，專注的在吃著，那麼我還是將我的手機放在褲袋內。

沐浴時間。原來永平寺是沒有私人浴室，大家都要根據指定的時間在公共浴室淋浴（當然是男女分開），我們一眾人排隊來到浴室門口，然後根據僧侶指示淋浴。打開門，是一個公眾浴室。我人生的首個公眾澡堂經驗就發生在永平寺，完全沒分你我，無遮無掩，赤裸裸的。

寺院 9 點關燈。我本來還在擔心會睡不著，也可能是日本和葡萄牙時差關係，我很快已經呼呼入睡了。

清晨 4 點起床，因為我們要先到大廳進行半小時的坐禪，然後參加於 5 時的早禱。看見全寺院超過 150 名僧侶一起頌經，心情絕對被這情景觸動，負責照顧我們的老師在頌經儀式過後帶我們參觀永平寺。才 7 時多，這裡的一切是那麼的平靜、祥和。

It's my first time practicing *Zazen* and staying in a temple. *Zazen* is like a meditation but it takes a different form. Eihei-ji, the Temple of Eternal Peace, is one of Soto Zen's two head temples. It is located deep in the mountains of Japan, near the city of Fukui.

Eihei-ji is huge, but we are not allowed to wander around. I have been given a schedule with everything planned, including my bathing time. Is there a private bathroom? No one dares to ask. Or maybe this is not really important because everyone come to experience zazen, and to seek inner peace. For me, I simply want to experience something different, and to learn slow living.

Dinner is served in a hall where every participant needs to follow certain rules: No one is allowed to talk, and we have to eat "not too fast," but "not too slow." And there's a proper ritual to keep our chopsticks after the meal. The vegetarian dishes, some using beans and tofu, are cooked perfectly, and are really tasty. Nothing fancy, but the original taste of the food comes through. How I wish I can take some photos, but no one seems to be doing that. And the dining hall is so quiet. Did I mention that no one is talking?

And then it's bath time. We all line up and are brought to the bathing place. And then I find out: It is a public bathroom with no partitions. It looks like the bathhouses that I've seen in Japanese TV programs. My hosting monk explains the ritual, says "Enjoy your bath," and leaves. The rest of us all look a bit embarrassed, but I take off all my clothes in seconds, find myself a spot, and start bathing. It is the quickest bath I have ever had. I think I am done in 3 minutes!

I go back to my room. Lights out at 9 p.m. This is my first time in a temple—how can I sleep at 9 p.m.? LOL. And then I start feeling hungry too because I had dinner at 5:30 p.m. But I do eventually fall asleep somehow (perhaps it's the jet lag after coming back from Europe).

I wake up at 4 a.m. in order to attend the morning service at 4:30. It's a very solemn ceremony with over 100 monks chanting in the Great Hall. I am moved by the ambience and many thoughts come to my mind.

Ricotta 芝士凍豆腐
Tofu with Ricotta Cheese

在廟宇渡過一宵時，其中有一道豆腐菜式令我印象難忘，也激發了我的創作靈感，做出帶點西方特色的版本。Ricotta 芝士與豆腐很配，兩者的質感和味道相近，我用了帶甜的汁令味道更平衡，我稱之為陰陽。

One of the foods I enjoyed during my temple stay was tofu, which prompted me to create my own tofu dish with a Western touch. Ricotta cheese and tofu can go together very well indeed! Both are quite similar in taste and texture, and my sharp-sweet sauce provides a good balance to them both. I like to call this the yin and yang.

材料（4人份量）

絹豆腐	300克，冷藏至凍
日本豉油	50毫升
味醂	1湯匙
糖	2茶匙
鮮磨薑蓉	10克
Ricotta芝士	130克
新鮮羅勒	1把，切絲
木魚乾	適量（裝飾用）
烘香芝麻	適量（裝飾用）

Ingredient (Makes 4 servings)

Silken tofu, well chilled	300g
Japanese soy sauce	50ml
Mirin	1 tablespoon
Sugar	2 teaspoons
Grated fresh ginger	10g
Ricotta cheese	130g
Fresh basil	handful
Bonito flakes	for garnish
Toasted sesame seeds	for garnish

做法 How to cook

將豆腐完全瀝乾，放碟上。

預備汁料，將日本豉油、味醂和糖放小鍋內拌勻，以小火加熱3至4分鐘至糖溶。加入薑蓉，離火備用。

將芝士放豆腐上，灑上切成絲的羅勒、木魚乾和芝麻，倒入暖的汁料即時享用。

Drain the tofu thoroughly and place on a serving plate.

To prepare the sauce, lightly mix the soy sauce, mirin, and sugar in a small saucepan. Heat over low heat for 3 to 4 minutes, until the sugar has dissolved. Add the grated ginger and remove the saucepan from the heat.

Place the ricotta on top of the tofu and sprinkle with shredded basil, bonito flakes, and sesame seeds. Pour over the warm sauce and serve immediately. Enjoy!

永平寺（Eiheiji）

📞 (+81) 776 63 3102　　@ www.aihonzan-eiheiji.com

京都 Solaria Nishitetsu Hotel Kyoto Premier

📞 (+81) 757 08 5757　　@ www.solaria-kyoto.nishitetsu-hotels.com

加賀 Hoshino Resorts KAI Kaga

📞 (+81) 503 78 61144　　@ www.kai-ryokan.jp

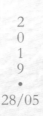

電視遙控器　Thoughts on the TV Remote Control

我是權力遊戲（Game Of Thrones）的超級粉絲。在葡萄牙和日本都未有時間看最新的劇情，終於今晚可以舒舒服服的在 TV 睇番之前預錄的集數。

拿著電視遙控器的按鈕在「撳、撳、撳」的時候，想起「撳、撳、撳」這個小時候被喻為極之神聖任務的主持人便是我的「媽媽」。

媽媽對我和妹妹的管教也頗為 strict，在 70、80 年代看電視也視作為一種 luxury item，所以我們只會在做完功課後才可以看電視，而看哪一個頻道的節目當然是媽媽的決定啊。

媽媽的控制權不單在電視遙控器，還有在廚房內。我們每天吃甚麼都是媽媽決定。現在人大了才會明白她的「control freak」，其實是她想 give the best to the family。這種心意在年青的時候我們還是不會明白的。

現在到媽媽的家吃飯，晚飯過後我們都會在客廳裡，電視在播住節目，但那個遙控器已經不會成為我虎視眈眈的目標了，因為我們各自已經有了智能電話，電視節目可以直接在我的電話內收看。媽媽說：「你們轉台吧，我沒有甚麼要看的 TV program」。

時代變了，我們的生活方式也變了，是好？是壞？我們再不用爭著用電視遙控器，因為我們看電視的模式已經變為 on demand，而我們每人已經有一部或以上的個人智能電話，這部電話也成為我們貼身的 entertainment device。大家有自己的私人空間，這不是很好嗎？

這個電視遙控器突然令我有感而發，想起小時候的軼事，當中也想起媽媽對家人的照顧和她對做菜的認真和執著。這份以往被我視為固執的性格其實就等同我現在對烹飪的一份熱情。

Like mother, like son，願我能夠繼續將這種熱情延續下去吧。

I need to catch up with all the *Game of Thrones* episodes I missed when traveling in Portugal and Japan for weeks. As I was using the TV remote controller to upload the missing episodes, I suddenly recalled a scene from childhood with my sister and I sitting in front of the television, with Mama holding the remote in her hand and pressing the buttons.

Yes, the remote control is the symbol of authority: Whoever holds the device is the person in control. And in my family, it was Mama. By the same token, Mama and Avo (Grandma) were the ones in control of the kitchen.

I remember Mama bustling around the kitchen, making sure my sister and I ate properly, and that Daddy got his dessert and coffee after dinner. And she spent hours and hours planning the family meals, and making sure every guest was happy and entertained. Her signature dish, Portuguese roast pig, was so crispy and tender! She knew that, and felt so proud to share her cooking with others.

嫲嫲和媽媽攝於中環擺花街金獅餅店外
Avo and Mama at Central

Mama's intention to be in control of the TV at home and her passion for cooking was simply her belief in sharing. And I am living on my own, and of course in full control of my own TV. But I also want to live like my mama and avo, sharing the joyous moments of cooking and making sure everyone is smiling and happy. I am so glad I am in full control of what I love to do!

薄荷蜂蜜紅蘿蔔沙律
Carrot Salad with Mint and Honey

這沙律本身也足以成為一道菜式，甚至是看電影的小食。這沙律宜室溫享用，可帶回辦公室成為健康午餐。

This is a delicious salad that can be a stand-alone dish or even a snack while watching your favorite movie at home. It's great at room temperature, so you can pack leftovers as a healthy lunch to enjoy at the office.

材料（4人份量）		Ingredient (Makes 4 servings)	
紅蘿蔔	700克，修好形狀	Carrots, trimmed	700g
蒜頭	2瓣，切碎	Garlic, chopped	2 cloves
蜜糖	30毫升	Honey	30ml
白酒醋	15毫升	White wine vinegar	15ml
初榨橄欖油	30毫升	Extra virgin olive oil	30ml
新鮮薄荷葉	1湯匙，切絲	Fresh mint leaves, shredded	1 tablespoon
鹽和白胡椒粉	適量	Salt and white pepper	to taste

做法 How to cook

預熱焗爐至180°C。在大焗爐盤內澆上大量的橄欖油，約125毫升，然後放入洋蔥、番茄、香葉和蒜頭在大湯鍋中加水至7至8分滿，加熱至水滾。將紅蘿蔔去皮，直切半，然後切成1厘米厚的長方形或半月形。

將紅蘿蔔放入鍋中煮8分鐘至軟，盛起瀝乾，稍放涼。放暖的紅蘿蔔放入大碗中，加入蒜、蜜糖、醋、橄欖油及薄荷，將所有材料拌勻，以鹽和胡椒粉調味即可。

Fill a large saucepan at least three-quarters full of water and bring to a boil. Peel and halve the carrots lengthwise, then chop into rectangular or half-moon shapes, about 1cm thick.

Drop the carrots into the boiling water and cook until just tender, around 8 minutes. Drain well and let cool slightly. Place the warm carrots in a large bowl and add the garlic, honey, vinegar, olive oil, and mint. Mix everything together and season to taste with salt and pepper.

無奈　Disoriented

這幾天都好像有點 disoriented，失去了在 social media 貼相的動力。

今天到 Towngas 爲明天的乳豬班做準備工作，要早一點出門，因爲港島那邊的交通有點緊張啊。

塞車。沒有問題啊。

工作完成了，回家路上，地鐵車廂大家都在看著自己的手機。到達了金鐘站轉車，無奈。

I don't feel quite right today
I don't have motivation
to take photos
to post photos
But I keep scrolling the newsfeed.

I don't feel quite right today.
I just want to go home
after class.

My train arrived at Admiralty
The platform is quiet
It must be busy up there.

On the train.
It's packed.
It's quiet.

I don't feel quite right today
I don't know why

Or …

It's disorienting.

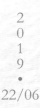
天啊，好熱啊！　　OMG, It's 32°C Here!

哎呀，還有不夠 1 個月的時間便要搬家了，I will miss this place for sure。

搬屋是一個掉東西的好機會，不過也是一個去 shopping 買東西的 good excuse。我走到皇后大道東，希望可以搜羅一張中式椅子，但始終還是敵不過這種悶熱的天氣！

突然想起媽媽在夏天會給我準備的涼粉，還有她特別用薄荷葉做的糖漿，再配上涼透心的士多啤梨或雲呢拿味雪糕，而媽媽總喜歡在夏日炎炎的日子為我們煮咖喱，因為她說咖喱有助去濕熱，對身體有幫助。我問媽媽：「那麼冬天要吃甚麼去驅濕熱呢」？

一直在想跟媽媽的笑話，心裡也笑了起來。想一想，不如做一大瓶水果酒 Sangria 吧！！

It's less than a month before moving to my new home! I will miss this place but I need something bigger—not fancy though—for my cooking. Yes, I am getting a NEW kitchen!

I went shopping today. I walked along Queen's Road East looking for a nice Chinese bench, but no luck at all. Every design was simply too "conventional" and didn't blend in with the rest of my furniture. Well, well, well… I enjoy furniture shopping, but not really on this day when the outside temperature is 32°C!

In this weather, I want a glass of sangria, or maybe a jar of sangria! Oh yes, and I remembered how, on hot days, Mama made the Chinese herbal grass jelly that she diced into little cubes and mixed with syrup and chilled in the fridge for hours. She served it with a scoop of strawberry or vanilla ice cream. Heaven! And on hot summer nights, she would cook curry for dinner. She said curry makes you sweat, and the sweat will carry the "wet and moist chi" from your body that makes you feel tired in the summer.

"How about winter, Mama? Do we need to get rid of our wet and moist chi?" I asked. And we all laughed.

水果酒
Sangria

水果酒 Sangria 來自西班牙，亦是葡萄牙常見的飲品。在夏末時的戶外聚會，以水果酒伴葡萄牙點心就最「正」。我家喜歡在夏天時做水果酒，但這個飲品當然是「成人 only」！

Sangria comes from Spain and is a common drink throughout Portugal, especially ideal for late summer al fresco brunches, and a perfect complement to a selection of *petiscos*. My family loves to make sangria during summer—kids allowed only after they turn 18 years old!

材料（一大瓶，6人份量）		Ingredient (Makes 1 jar, enough for a family of 6)	
紅酒	750毫升	Red wine	750ml
砵酒	80毫升	Port wine	80ml
白蘭地或威士忌	60毫升	Brandy or whiskey	60ml
七喜或雪碧汽水	350毫升	7UP or Sprite	350ml
橙味梳打	330毫升	Orange soda	330ml
蘋果	100克，切片	Apple, sliced	100g
橙	100克，切片	Orange, sliced	100g
士多啤梨	100克，切片	Strawberries, sliced	100g
黃糖	50克	Brown sugar	50g
肉桂條	1條	Cinnamon stick	1
八角	3粒	Star anise	3 pods

做法 How to cook

將所有材料放入一個大瓶子內，然後放雪櫃冷藏至少4小時，或將瓶子放在鋪滿冰塊的大碗內冷凍。

Combine all the ingredients in a large jar and refrigerate for at least 4 hours, or let it rest in a cool place in a large bowl of ice cubes.

你可用白酒代替紅酒，及用藍莓，做成水果白酒。

You can make white sangria by using white wine instead of red and adding blueberries and strawberries.

蟹肉咖喱
Flaked Crabmeat Curry

這是我家的食譜，媽媽會在夏天做這個咖喱，叫我們吃得全身冒汗。我最喜歡咖喱中的小雞蛋（當時我不知道它們是鵪鶉蛋）。我和妹妹 Karen 吃這道菜時會先玩一個遊戲，先將 5 隻小雞蛋放入口中，然後講故事，但不可以咬到蛋，這就是我們的玩意！

This is a family recipe and also the curry that Mama made during the summer to make us sweat. The thing that I love about the curry is the "mini eggs" (I had no idea those were quail eggs at that time). Karen, my younger sister, and I loved to play a game by putting five mini eggs into our mouths, and then we had to tell a story without biting the mini eggs.

材料（4人份量）		Ingredient (Makes 4 servings)	
橄欖油	1湯匙	Olive oil	1 tablespoon
無鹽牛油	80克	Unsalted butter	80g
洋蔥	1個，切碎	White onion, chopped	1
蒜頭	2瓣，切碎	Garlic cloves, chopped	2
紅椒碎	2條，切幼碎	Red chilies, finely chopped	2
咖喱粉	2湯匙	Curry powder	2 tablespoons
麵粉	25克	All-purpose flour	25g
月桂葉	2片	Bay leaves	2
雞湯	500毫升	Chicken broth	500ml
蟹肉	500克，拆成小片	Crabmeat, flaked	500g
蝦	200克，蒸熟去殼	Shrimp, steamed and shelled	200g
椰奶	100毫升	Coconut milk	100ml
鵪鶉蛋	12隻，焓熟去殼	Quail eggs, cooked and peeled	12
鹽和白胡椒粉	適量	Salt and white pepper	to taste

做法 How to cook

在炒鑊中以中火燒熱橄欖油，放入牛油，再放入洋蔥炒至軟身，加入蒜和辣椒再炒2分鐘至散發辣味。

調至小火，放入咖喱粉、麵粉和月桂葉炒勻，逐少倒入雞湯，不斷攪拌至煮成略濃稠的湯汁，約需8分鐘。

再以小火煮5分鐘，不時攪拌，然後加入蟹肉和蝦肉炒數分鐘，至所有材料混合。

加入椰奶和鵪鶉蛋，以鹽和胡椒粉調味，調至中火煮3分鐘至所有材料熱透，棄去月桂葉即可享用。

In a sauté pan, heat the olive oil over medium heat. Add the butter and then the onion and sauté until soft. Add the garlic and chilies and sauté for 2 minutes, until you can smell the spicy aroma from the chilies.

Adjust the heat to low, stir in the curry powder and flour, and add the bay leaves. Slowly add the broth, stirring constantly, until the sauce becomes slightly thick, about 8 minutes.

Simmer over low heat for 5 minutes, stirring from time to time. Add the crabmeat and shrimp, stir, and simmer for a couple of minutes, until all the ingredients are well blended.

Add the coconut milk and quail eggs and season with salt and pepper. Bring back to a simmer over medium heat and cook until everything is heated through, about 3 minutes. Discard the bay leaves before serving.

忙　Busy

　　痴線喋！6 月好像一瞬間便過去了。在葡萄牙和日本回來後便一直的忙著工作。慶幸品牌書《微壓潮煮》已經完成，準備在 7 月的書展中跟各位見面。

　　有時我問自己：值得嗎？下一步呢？

It's been a very hectic June. I have been working on two new books simultaneously and excited that my second book, *Woll Low Pressure Cooker Recipes*, is completed and ready for publication at the HK Book Fair 2019 in just a couple of weeks. Looking at my calendar for the coming months, it's so packed.

Sometimes I do ask myself: Is it worth it? What is next?

I need to cook something today to clear my mind and give myself a little break. I want to do something sweet. Okay, how about something that requires some "digging"?

日本南瓜蒸椰子布甸
Coconut Custard in Japanese Pumpkin

椰子和南瓜很配，可煮鹹食如咖喱，也可一起做成簡易的甜品，人人都會愛！

Coconut and pumpkin always work together on good terms. You can use them in savory dishes like curry, but they also combine for a very simple dessert that will impress everyone.

材料（4人份量）		Ingredient (Makes 4 servings)	
日本南瓜	1個，約12厘米直徑	Japanese pumpkin	1 (about 12cm in diameter)
雞蛋	8隻，拂勻	Eggs, beaten	8
椰漿	200毫升	Coconut cream	200ml
椰糖或黃糖	150克	Palm sugar or brown sugar	150g
班蘭葉	4塊，切絲	Pandan leaves, shredded	4
鹽	½茶匙	Salt	½ teaspoon

做法 How to cook

南瓜洗淨抹乾，橫切去頂部，切出約9厘米直徑的南瓜蓋，用湯匙刮出南瓜籽棄掉，南瓜備用。

在大碗中打入雞蛋和椰漿拌勻，加糖、班蘭葉和鹽。我會用手榨壓班蘭葉至軟身及釋放香氣，並拌至糖完全溶化。將混合物過篩，倒入南瓜盅內至快到頂，蓋上南瓜蓋。

小心將南瓜放入大蒸鍋中，鑊至少要比南瓜大一倍。蓋好以中火蒸30至40分鐘，至混合物凝固熟透及南瓜軟身。稍涼後整個南瓜連南瓜蓋上碟享用。

Wash and dry the pumpkin. With a sharp knife, remove the top of the pumpkin by cutting around the top to create a "lid" about 9cm in diameter.. Scoop out all the seeds from the pumpkin with a spoon and discard. Set the pumpkin and its lid aside.

In a mixing bowl, whisk the eggs and coconut cream together, then add the sugar, pandan leaves, and salt. I use my hand to squeeze the mixture until the pandan leaves are soft and aromatic and the sugar has dissolved completely.

Strain the mixture through a strainer into the pumpkin, filling to just below the top. Place the pumpkin lid on top. Carefully place the pumpkin in a large steamer, or a wok at least twice the size of pumpkin. Cover and steam over medium heat for 30 to 40 minutes, until the custard is firm and cooked and the pumpkin is soft.

Let cool slightly, then serve whole with the reserved lid on top.

安樂窩　Home Sweet Home

不知不覺已經搬到新居接近 2 星期了，鹹蝦燦新書拍攝工作也於昨天開始。之前放了幾張新廚房的相片在 Facebook 和 Instagram，粉絲們都大讚特讚。

甚麼是安樂窩？說句老實話，我真的不用睡一張超級勁大 king-sized bed，也不用住一間偌大的 apartment，一間簡單、舒服和給我有安全感的 apartment 已經是我的安樂窩。

每天早上陽光從窗外射進來，然後自自然然的睡醒，心裡已經急不及待的為今天的事情去忙，做自己喜歡的事情，這種感覺是非常幸福的。回想起自己在很多年前創立的公司，曾經有一段時間遇上很多的壓力，晚上失眠，第二天早上不願醒來，有時更有奇想希望今天可以停頓下來。對啊！是成是敗轉為空，幾度夕陽紅。

過去的幾個星期很多朋友可能對「安樂窩」有著很多的體驗和感受。我非常愛我的「安樂窩」，我的家。但願無奈、不安和無助的感覺可以快些過去。

爸爸和嫲嫲，你倆在守護著我們一家人的。

My home does not need to be stylish to impress others. My home does not need to be so huge that I could host a party for 50 people. My home does not need to be equipped with the state-of-the-art tools for me to cook. My home only needs to be a place that I feel warm and safe.

I moved into my new home a couple of weeks ago and I am glad things are settling down. My dogs are happy, and excited to explore the new areas of this part of the island, and my helper has met a few new friends during the dog walking. My new kitchen is running smoothly, my new SMEG oven is good; I made Neo Molotoff (*The Ham Har Chaan Cookbook*, p. 265) last week and it came out perfect.

And I sleep like a baby in my new bedroom. Do you love the moments of waking up in the morning? The feeling of looking forward and achieving something? A few years ago, there were times that I was afraid to wake up in the morning because of the stress from work and other problems that were troubling me. That

was a very tough time of my life. I am glad my family stood by me. The older I get, the more I believe in the saying that family is inseparable.

I am so lucky to have a home sweet home. Of course there are still many things that I want to achieve, but I am happy, now, for what I have and who I am.

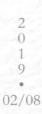

Facebook, Instagram, YouTube

記得在最初踏入「烹飪界」圈子，認識了一位名 blogger，還是非常「innocent」的我曾經膽大大的請那位前輩「關照小弟」，希望可以出席一些品牌宣傳活動（event），又或是介紹品牌的產品。那位 blogger 笑笑口的跟我說：「你的粉絲人數這麼少，恐怕有點難啊！你不如有再多些粉絲人數時再跟我說吧」。這句說話我永遠的記在腦海裡。

從那一刻開始，我便知道原來粉絲人數多一點就會「有 say」、「有品牌」，不過我始終相信實力和 strategy 才是最重要，所以我便很努力的在「烹飪界」打拚，每天學習、練習和爲自己的「烹飪事業」重新部署，給自己寫了一個 plan，哈哈，bingo！

在 Facebook 拼了接近 3 年，開始也發展其他平台，例如 IG 和 YouTube，但真的 OMG，這 3 個平台的策略都絕對不同，我頓時要做下 3 個完全不一樣的市場策略。

既然我打算將我的日誌放在這書裡，就容許我用以下的文字和空間賣些自我宣傳廣告吧：

各位讀者，大家好！我是 John Rocha，我不是一位 blogger，也不是一位網紅，我是一位 micro influencer，請各位多多賜教。歡迎各位的關注和追蹤我以下的平台：

f eufoodjourney

🅾 eurasianfoodjourney

▶ 鹹蝦燦 The HamHarChaan

Are these the rules of the social media game?

Less than 10 Likes = Bad
20 Likes = OK
30 Likes = Good
40 Likes = Better
50 Likes = Well done
60+ Likes = Heaven

More important: Is this how to get more recognition?

I've heard that you should post a nice recipe photo on Facebook, and ask people to check the LIKE button, and then key in certain default words in the comment box—in return they get a free recipe! If you do that, then the number of LIKES and Comments in that particular post will go all the way up to over 1,000!!!!

Sorry. This is not my game. If I want to share my recipes on social media, I will simply put them up with no other conditions. You can buy my cookbook, and sign up for one of my cooking classes if you want to know more. Yes, come to my cooking classes I can assure you they are very interactive, and also just so much fun.

And if you want to know how to make the African chicken and pasta with mozzarella and tomatoes? Check my social media and you'll find videos.

Okay, all that said, it's time for some self-promotion: Please follow my Facebook page and Instagram, and subscribe to my YouTube Channel. I am John Rocha, a cook, a food writer, and a social media host.

非洲雞
African Chicken

非洲雞不是來自非洲，也不是來自葡萄牙，它有點像巴西的香辣烤雞、印度的青辣椒咖喱雞和葡萄牙的霹靂霹靂烤雞。非洲雞的食譜源自 1940 年代末，一位澳門廚師 Americo Angelo 到訪莫桑比克，發現霹靂霹靂烤雞 —— 一種佈滿霹靂霹靂辣椒醬汁的辣味烤雞。Americo 將這靈感帶回澳門，創作出自己的版本，在香濃醬汁中加入椰子味道。

African chicken isn't from Africa, nor is it from Portugal. However, it is similar to *frango assado* from Brazil, chicken *cafreal* from Goa, and piri piri chicken from Portugal. The original recipe was inspired by Macanese chef Americo Angelo's visit to Mozambique in the late 1940s, where he discovered piri piri chicken, the spicy grilled chicken dripping with a sauce made with piri piri (aka peri peri). Americo brought this inspiration back to Macau and created a version of his own with a coconut-rich sauce.

材料（4人份量） / Ingredient (Makes 4 servings)

醃雞 / Chicken and Marinade

蒜頭	8瓣，切蓉	Garlic, minced	8 cloves
乾蔥	8粒，切蓉	Shallots, minced	4
紅辣椒	1湯匙，連籽切粒	Red chili, diced with seeds	1 tablespoon
豉油	2茶匙	Soy sauce	2 teaspoons
鹽	2茶匙	Salt	2 teaspoons
糖	1茶匙	Sugar	1 teaspoon
白胡椒粉	1茶匙	White pepper	1 teaspoon
雞腿	4隻（250克）	Chicken thighs with bones	4 (250g)
橄欖油	1湯匙	Olive oil	1 tablespoon

醬汁 / Sauce

蒜頭	4瓣，切蓉	Garlic, minced	4 cloves
乾蔥	2粒，切蓉	Shallots, minced	2
紅辣椒	2條，切片	Red chilies, sliced	2
茄膏	3湯匙	Tomato paste	3 tablespoons
茄汁	2湯匙	Ketchup	2 tablespoons
雞湯	300毫升	Chicken broth	300ml
花生醬	1至2湯匙	Peanut butter	1 to 2 tablespoons
椰奶	100毫升	Coconut milk	100ml
辣椒仔（Tabasco）或霹靂霹靂（piri piri）辣汁	2茶匙	Tabasco or piri piri sauce	2 teaspoons
椰絲	2湯匙	Desiccated coconut	2 tablespoons

做法 How to cook

預備雞腿：在大碗內將蒜蓉、乾蔥蓉、辣椒、豉油、鹽、糖和胡椒粉拌勻，然後放入雞腿醃勻，蓋好放雪櫃冷藏過夜（至少12小時）。

煎鑊內倒入橄欖油，以中火加熱。將雞腿瀝乾，抹去醃料備用。將雞腿放入鑊中煎至金黃香脆及熟透（見tips），約需8分鐘，期間需反轉，將雞腿盛起備用。

預備醬汁：原煎鑊放入醃汁中的蒜蓉、乾蔥蓉，再加入醬汁材料的新鮮蒜蓉、乾蔥蓉和辣椒，以小火炒1分鐘。加入茄膏、茄汁和雞湯，以小火邊煮邊攪拌3分鐘。加入花生醬和椰奶再煮1分鐘，至湯汁濃稠香滑，最後加入辣汁調味。

將雞腿放焗盤上，淋上醬汁，灑上椰絲。

將焗爐調至燒烤模式（200°C），放入雞腿烤4分鐘至表面金黃即成。

For the chicken and marinade: In a large bowl, mix together the garlic, shallots, chili, soy sauce, salt, sugar, and pepper. Add the chicken and turn to coat all over. Cover and marinate in the refrigerator overnight (at least 12 hours).

Heat the olive oil in a frying pan over medium heat. Scrape the garlic, shallots, and any liquid off the chicken and set aside in a small bowl for the sauce. Add the chicken to the hot oil and cook, turning, until the skin is brown and crispy and the chicken is cooked through (see Tip), about 8 minutes. Transfer the chicken to a plate and set aside.

For the sauce: In the same frying pan, combine the garlic and shallots from the marinade with the fresh garlic, shallots, and chilies and sauté over low heat for 1 minute. Add the tomato paste, ketchup, and broth. Cook for 3 minutes over low heat and keep stirring. Add the peanut butter and coconut milk and cook for 1 minute longer, until the sauce is slightly thick and creamy. Add the Tabasco or piri piri at the end.

Place the chicken thighs on a baking sheet, pour the sauce over, and top with the coconut.

Turn your broiler to high (200°C). Put the baking sheet under the broiler for 4 minutes, or until the surface of the chicken is browned.

視頻食譜
Video Tutorial

要知雞腿是否煮熟，可用小刀插入最厚肉的部位，
開口不用太大，然後用刀叉拉開看看肉的顏色，
如是白色即熟透，如仍帶粉紅則需再多煮一會。

To check if the chicken is cooked through, make a small incision in
the thickest part of the chicken and check the color.
If you don't want to cut apart your chicken, a small incision will work.
Just pull apart the sides using a fork and knife
until you can see the color of the meat all the way through.
If you see the meat with pink hues in the white,
it has to cook longer. If the meat is white, then the chicken is ready.

芝士車厘茄金不換長通粉

Penne with Mozzarella, Cherry Tomatoes, and Thai Basil

這個菜式很易做，它亦也很適合配燒烤食物一同享用。金不換為這意粉帶來泰式風味，而且它比羅勒甜一些，帶少許甘味，可在超市或街市買到。

This is an easy pasta and great with barbecue too. I like to add a Thai touch by using Thai sweet basil, which has a sweeter flavor than regular basil, with a hint of licorice. You can easily find Thai basil at the supermarket or fresh market.

材料（4人份量）		Ingredient (Makes 4 servings)	
初榨橄欖油	3湯匙	Extra virgin olive oil	3 tablespoons
蒜頭	2瓣，切片	Garlic, sliced	2 cloves
車厘茄	300克，切半	Cherry tomatoes, halved	300g
鹽和胡椒粉	適量	Salt and pepper	to taste
坑紋長通粉	350克	Penne rigate	350g
新鮮金不換葉	10片，切絲	Fresh Thai sweet basil, shredded	10
水牛芝士球	125克，切1厘米方粒	Mozzarella balls, cut into 1cm cubes	125g
青檸汁	2茶匙	Lime juice	2 teaspoons

做法 How to cook

大煎鑊中倒入油，以中火加熱，放入蒜頭和車厘茄，輕力炒1分鐘。大湯鍋中煮滾水，加適量鹽，放入長通粉煮至軟硬適中，盛起放入步驟1的煎鑊中，以中火加熱。將金不換、水牛芝士放入煎鑊中，以鹽和胡椒粉調味，離火，加入青檸汁拌15秒至所有味道混合，即可享用。

Heat the oil in a large frying pan over medium heat. Add the garlic and cherry tomatoes and gently cook for 1 minute, stirring. Remove from the heat, and set aside. Bring a large saucepan of salted water to the boil and cook the pasta until al dente. Drain and tip the pasta back into the same frying pan.

Add the Thai basil and mozzarella to the pan. Season with salt and pepper. Remove from the heat, add the lime juice, and toss everything together for 15 seconds to ensure all the flavors combine.

初榨橄欖油的冒煙點較其他油低（190°C），所以也適合做以小火烹調的清爽菜式。如你喜歡辣味，可在煮好的長通粉上灑上指天椒粒，甚至加入與蒜片同炒。

Extra virgin olive oil has a lower smoke point (190°C) than other oils, and it is safe to use for light, lower-heat cooking.

If you want some kick, add some Thai bird's eye chilies to the finished pasta just before serving or when you sauté the garlic.

視頻食譜
Video Tutorial

家電 · 家品 · 博覽 2019　Home Delights Expo 2019

　　記得有同行朋友跟我說能夠被公司獲邀參加一年一度的博覽烹飪示範猶如中獎，是一件值得高興的事。很多大品牌都會利用這機會在找合適的 KOL、influencer 和廚師合作。我非常幸運可以連續 3 年繼續在 Home Delights 參與廚藝表演的環節。希望繼續受到別人的賞識啊！

It's already my 3rd year joining Home Delights. I see every show as my last show, and this is my drive to make my cooking show "the best of the best."

今天是我的牛一　　Happy Birthday

今天是我的生日，祝自己生日快樂！

今天休息一天！

August 28 is my birthday! No cooking today.

慢活　Slow Living

在曼谷私穩度高的 The Siam 住了 2 晚，今天早上起來便出酒店逛一逛，吸吸曼谷清晨的空氣。我喜歡大清早聽著歌，然後在街上走走的感覺。

本想來曼谷 take a little break，但心裡始終是放不低不同的 project。這本書接近「埋尾」的階段了，我寫的東西會吸引嗎？出版商滿意嗎？料理班的安排如何？還有 10 月清邁的美食之旅。

在我前面出現了一位應該在化緣的和尚。嗯，泰國的僧侶每天大清早都會光著腳的去化緣，他經過我身旁時，我倆眼神互相望一望，然後我很自然的跟僧侶說一聲：sawadee，那名僧侶竟然停下來，然後用英文問我：「where from?」

「Hong Kong」我說。

「Very good」他說。

僧侶望著我，再說：「Shoe not ok, slow living, you try」。

幾秒鐘的對話，卻令我不停在想這句話的意思，難道他是指我走得太快，要慢下來的意思？又或是像他光著腳慢慢的走，感受慢活的意義？

晚上跟我在曼谷好友聚舊，跟他說了今天與僧侶的「半分鐘對話」，他聽完後立即說：「他的意思是叫你去做一些 donation，幫助有需要的人啊！」然後我們的話題又回到曼谷 5 大出色餐廳上。

回酒店的路上，我不停反思，給自己一些空間去停一停，想一想。很快這本書便要誕生了，也意味著這個 project 也快完結了。So what is next？

I got up pretty early this morning here in Bangkok because I wanted to go for a run around my hotel, the Siam. It's only 6 a.m., so the streets are not congested at all. I love exercising early in the morning, turning on my playlist from iPhone, and enjoying the quiet and peaceful morning on the streets of Bangkok.

I must admit my mind is not focusing on the music as I keep thinking about the gastronomy trip to Chiang Mai, Thailand, in a few weeks' time. Also on my mind: my cooking class at Towngas next week, and the closing chapter for my book, which I plan to write during this short trip to Thailand as my post-birthday celebration.

At daybreak in the Dusit area, a Buddhist monk in saffron robe appears barefoot in front of me. He must be on his way to collect alms in the neighborhood. Out of curiosity, I use my phone and take a few snapshots as he walks towards me. What am I thinking? Maybe, to put the images into my image bank for my book?

As the monk walks pass me, we smile at each other and then I said "Sawadee." He stops and then smiles, asking me in English "Where from?"

"Hong Kong," I said.

"Very good," he replied.

The monk looked at me. Then he said, "Shoe not okay, slow living. You try."

And then he continued to walk pass me, heading in the direction that I came from.

What is he trying to say? Is he saying I should slow down? Is he saying I should live like a monk and collect alms? It reminds me of my temple stay earlier in Japan.

Slow living. Yes, this is something that I always want to practice. Easy to say, but difficult to do. LOL. What am I seeking? Or is something seeking me?

My friend Nat and his wife invited me tonight for a dinner at their place. It was a simple and lovely dinner. I love the clams cooked in Thai basil and the Thai-style roast chicken is simply divine.

I was telling Nat and his wife about what the monk told me this morning. They said I should donate my sneakers and some money to help the needy. Oh yes, of course I should do it. And maybe a charity program to cook for the needy? And how about asking my cooking students to join me and cook for love to raise money to buy shoes for some needy children?

辣椒膏炒蜆
Clams with Roasted Chili Paste

這是我最喜歡的泰菜之一。我記得多年前跟爸媽到泰國旅遊，在 Sukhumvit Soi 24 的海鮮市場餐廳內吃這道炒蜆，爸爸還用饅頭點汁吃。這食譜很容易，就算新手也輕易做得到。

This is one of my all-time-favorite Thai dishes. I remember my parents loved to order it whenever we visited Seafood Market Restaurant at Sukhumvit Soi 24, many years ago. Papa enjoyed dipping Chinese fried plain bun into the sauce. The recipe is so easy that even one without much cooking experience can handle it with no problem.

材料（4人份量）

菜油	2湯匙
蒜頭	3瓣，切碎
鮮蜆	500克
魚露	1湯匙
糖	1茶匙
泰式辣椒膏	2湯匙
紅辣椒	2條，切片
金不換	2把

Ingredient (Makes 4 servings)

Vegetable oil	2 tablespoons
Garlic, chopped	3 cloves
Fresh clams	500g
Fish sauce	1 tablespoon
Sugar	1 teaspoon
Roasted chili paste	2 tablespoons
Red chili, sliced	2
Thai basil	2 handfuls

做法 How to cook

以中火燒熱中式鍋或大炒鑊，放入蒜碎和鮮蜆，炒至蜆開口。

加入魚露、糖、辣椒膏再炒2分鐘至醬汁和蜆肉混合。加入切片的辣椒和金不換炒勻，棄去不開口的蜆，趁熱享用。（我愛用手食用，哈哈）

Heat the oil in a Chinese wok or large sauté pan over medium heat. Add the garlic and clams at the same time and stir-fry until the clams begin to open.

Add the fish sauce, sugar, and chili paste and stir-fry for 2 minutes, until the sauce blends with the clams. Add the sliced chili and Thai basil and stir to mix. Discard any clams that have not opened. Serve hot. (And I love to use my hands to eat them. Hahaha.)

烤雞
Grilled Chicken

對！我非常喜愛吃雞！這食譜可如平時一樣用焗爐烤焗，或者帶到戶外 BBQ，配糯米飯和蔬菜沙律最夾。

Yes, I do love chicken! Use your broiler for this one, like I normally do, or take it outside to the barbecue. Perfect with Thai glutinous rice and a green salad.

材料（2人份量）		Ingredient (Makes 2 servings)	
蒜肉	8粒，拍扁	Garlic, crushed and peeled	8 cloves
鹽	1茶匙	Salt	1 teaspoon
黑胡椒碎	1茶匙	Ground black pepper	1 teaspoon
椰糖或黃糖	1湯匙	Palm sugar or brown sugar	1 tablespoon
豉油	1湯匙	Soy sauce	1 tablespoon
雞	1隻約550-700克，沿脊骨切開壓平	Whole chicken (butterflied)	1 (550 to 700g)

做法 How to cook

在大碗內放入蒜、鹽、黑椒碎、糖和豉油拌勻。放入雞與醃料拌勻，按摩一會，蓋上保鮮紙，放雪櫃醃3小時。煮前30分鐘將雞從雪櫃取出放室溫回溫。如用炭烤，先將碳燒至白。如用氣體燒烤爐，先以150°C預熱。將雞以中低火烤40分鐘至熟透，中間反轉一次。你也可用對流電焗爐，將雞放架上，離熱源約10厘米，兩邊各烤20分鐘至熟。

In a large bowl, mix the garlic, salt, black pepper, sugar, and soy sauce together until thoroughly combined.

Add the chicken and massage with the seasoning mixture. Cover with plastic wrap and let marinate in the fridge for 3 hours. Let the chicken come to room temperature for 30 minutes before cooking. Prepare a charcoal grill with glowing hot coals, or preheat a gas grill to 150°C. Grill the chicken over medium-low heat, turning once, for 40 minutes, until cooked through. Alternatively, broil the chicken under a broiler rack about 10cm away from the heat source for about 20 minutes on each side, until cooked through.

要知道雞是否煮熟，將刀插入最厚肉的地方，看看肉的顏色，切口不用太大，用刀和叉向兩邊拉開，如見雞肉白中帶粉紅，要多烤一會，如果雞肉全白，即已熟。

To check if the chicken is cooked through, make a small incision in the thickest part of the chicken and check the color. Just pull apart the sides using a fork and knife until you can see the color of the meat all the way through. If you see the meat with pink hues in the white, it has to cook longer. If the meat is white, the chicken is ready.

交稿了　It's the Due Date

出版商說不再等了，今天是截稿的最底線。不知不覺已經寫了很多篇的日誌，也煮了很多款的菜式。最近的 3 個月情緒也是起起伏伏，有起有跌、有無奈有憧憬的、有活崩崩有死去活來的、有充滿了意義有充滿了泛味的。無論是好是壞，只要能夠給我回到家，在廚房做起菜來，一切都是美好的。前幾天媽媽來到我的家，我在廚房做她喜歡的燒肋骨，看見她在廳看著我的書，吃飯時對我做的 spare ribs 大讚特讚，那一刻是滿足的。我把電視關掉，享受跟家人飯聚之樂，望著窗外遠處的香港島，我知道一切很快會安好的。

My publisher has given me a final ultimatum. LOL ! I need to submit final copy and give a green light to my new book.

I have written a lot of journal entries and recipes over the past 300 days. There are good times and bad times, days that I feel so hyper and days that I feel so moody, days that I want to achieve so much and days that I just want to do nothing. And no matter what, whenever I am back to my home sweet home, and spending time cooking in my kitchen --- well, those are the magic moments when I feel a peace of mind.

Mama came to my place for dinner few days ago. While I was preparing her favorite Portuguese spare ribs, I saw her looking at my books with a smile on her face. And, suddenly, I felt a strong sense of happiness and fulfillment. I turned off the TV news. I just want to treasure the moment with my family together.

I look out from my living room window and there's my beautiful home, Hong Kong, right across the harbor. And I know things will be better very soon.

葡式烤排骨
Portuguese Ribs

這是里斯本一位親戚的食譜，當時的情景仍然很鮮活：所有人坐在長枱的兩邊，枱上放滿食物和一籃籃的麵包，然後 Auntie Florinda 捧出一盤烤排骨，大家一起開動。Sweet！

This is a family recipe from relatives in Lisbon: I remember so well, everyone seated along a long bench with an array of food and baskets of bread. And then Aunt Florinda brought out a stack of ribs, and put them on the table — everyone just grabbed for them.

材料（4人份量）

蒜頭	6瓣，切碎
粗鹽或海鹽	1湯匙
黑胡椒粒	2茶匙
檸檬汁	3湯匙（1個檸檬份量）
紅辣椒	3條，切碎
新鮮羅勒	一把，切碎
豬肋排或豬仔骨	1副，約780克
乾性白酒或玫瑰紅酒	240毫升

Ingredient (Makes 4 servings)

Garlic, chopped	6 cloves
Coarse salt or sea salt	1 tablespoon
Black peppercorns	2 teaspoons
Lemon juice	3 tablespoons (1 lemon)
Red chilies, chopped	3
Fresh basil, chopped	handful
Rack of pork spare ribs or baby ribs	1 (about 780g)
Dry white or rosé wine	240ml

做法 How to cook

將蒜頭和鹽放入研缽內，研磨成醬，加入胡椒粒、檸檬汁、辣椒和羅勒，繼續研磨至混合。

將排骨放大碟或大碗內，均勻地抹上磨好的醬，倒入酒至蓋過排骨，如不夠可酌量再加。放雪櫃醃1小時，天冷時可放室溫。

以200°C預熱焗爐15至20分鐘。

將排骨瀝乾（醃料留用），放在焗架上，底墊一個有邊焗盤，放入焗爐焗30至40分鐘，期間不時掃上預留的醃汁，焗至排骨熟透軟腍。

Using a mortar and pestle, mash the garlic with the salt, forming a paste. Add the peppercorns, lemon juice, chilies, and basil, stirring to blend.

Place the ribs in a large dish or bowl and coat evenly with the seasoning mixture. Pour the wine over the ribs, adding more wine as necessary to coat all sides. Marinate for 1 hour in the refrigerator or at cool room temperature.

Preheat the oven to 200°C for 15 to 20 minutes.

Remove the ribs from the marinade (reserve the marinade). Place the ribs on a rack in a rimmed baking sheet and bake for about 30 to 40 minutes, brushing the ribs with the reserved marinade from time to time, until the ribs are cooked through and tender.

2018/07/09
我們出發了 Here We Come, Malacca!

難道我真的發神經？竟然在凌晨 4 時 30 分寫這篇 journal！

可能是緊張的關係，今天起得特別早，而再過多幾小時，我便會和一班料理班同學、一班粉絲飛到馬來西亞了。

是的，我喜歡旅行，喜歡住一些「正」、「型格」的酒店。

是的，我喜歡飛在 30000 呎的天空上，望出窗外，很喜歡看見窗下的一片雲海，而飛機引擎永遠在我的後面（哈哈，明白我的意思嗎？）

坐飛機，我的 favourite seat 是 1A、1K 或是 11A 和 11K，好了，好了，不要說我浮誇。我認識有位朋友花上很多錢買茶葉，難道這是浮誇？

喜歡紅酒的朋友更多的是，我這個喜歡「享受一點的旅遊方法」也沒有太大的問題啊！

就是這樣，有同學向我提出舉辦「John Sir 美食遊」，不是「趕頭趕命」的那種，而是我喜歡的那種。好的，我就叫這種旅遊為「elite travel」。

除了期待今次旅程的節目，我更期待是和同學們一起生活數天，大家一齊上料理班，一起吃喝玩樂，這絕對是一個 unforgettable experience。而對我來說，也是一個新的里程碑。

Ok 啦！我要往機場了！我們出發了！

It's 4:30 a.m. The flight for my first gastronomy tour of Malaysia leaves in 3 hours and I am all ready. I've done all the planning, I know the itinerary by heart, and I'm pretty sure everything will go smoothly.

I love to travel. I especially love to travel business and first class. To me, travel is a pampering, and I want to enjoy and relax to the max. Arriving in good shape can make a big difference.

I love to stay in boutique hotels or famous branded names that offer personalized service. I like it when hotel staff calls me by name, as it's always the home feeling that I look forward to.

I love doing nothing, staying in a hotel, enjoying my breakfast, and reading my favorite book by the hotel pool.

And I love getting a private tour guide and seeing things that are not in the travel books. I want to experience what a local does.

I love to spend hours in a supermarket looking for food produce, enlightening my curiosity, and learning new dishes from locals. And I love to shop for books about local delicacies.

It turns out quite a lot of my cooking students and fans enjoy this kind of travel lifestyle. And so I had this idea to organize a holiday filled with nice food, cooking, culture, nice hotels, and chillaxing time. Above all, it is the time that we all spend together that is the most cherished part.

Am I crazy to write a journal at this time of the
day? I just want to capture this feeling when I can.
We are going to have a fabulous time together. Here
we come, Malacca!

後記二 Afterword 2

2018/10/09

在馬六甲尋找鹹蝦燦　Kristang Cooking

　　決定以馬來西亞爲我首個「John Sir 美食遊」的地方，是因爲我希望每一個旅遊的城市都有「飲食文化」的元素。

　　馬來西亞的馬六甲跟葡萄牙也有一段很深的歷史淵源，應該是早於 16 世紀。葡萄牙人跟當地的馬拉人通婚，從而發展了獨有的馬葡料理，稱爲「Kristang」。Kristang 料理比娘惹菜出現得更早，也可以說是 fusion 菜的始祖。

　　我們來到了馬六甲的 Majestic Hotel，在這裡的餐廳吃了一頓非常美味的 Kristang 料理，同學們也覺得 Kristang 料理跟葡菜在食材、做法也頗爲相似。Well，大家都是鹹蝦燦啊！（註：甚麼叫「鹹蝦燦」呢？可以參看《鹹蝦燦之味》第 4 及 6 頁，有更詳盡介紹。）

　　我也非常高興認識到這裡的總廚 Chef Melba，她跟我介紹了 Kristang 料理中的招牌菜：焗釀蟹蓋和魔鬼咖喱。如果大家也想了解澳葡菜的焗釀蟹蓋，可以參考《鹹蝦燦之味》第 121 頁，而魔鬼咖喱則跟在 213 頁介紹的非洲雞很相似。

　　吉隆坡的 Banyan Tree Kuala Lumpur 實在太棒了！偌大的房間，亮點就是可以遙望到 KL 雙子塔的浴室，在這裡浸一個泡泡浴，望著外面的 city view，頓時會感嘆一句：「This is life!」寫到這裡，突然對明天「黃阿華燒雞翼」非常的期待啊！

Kristang are of Portuguese-Dutch descent and their roots can be traced back to Malacca, 2 hours' drive from Kuala Lumpur, Malaysia. Malacca had been influenced by Portuguese Empire from the 16th century. The Kristang community has a rich culinary heritage, developed over the centuries and influenced by other ethnic cuisines. Kristang cooking is very similar to Macanese cooking, as it is influenced by Portuguese, Dutch, and British cuisines, and it also has the local flavor of Malay, Chinese, and Indian cuisines.

The trip to Malaysia has been great so far. Everyone has been enjoying themselves and I have been so busy running around. The cooking class at Casa del Rio Melaka was interesting, as we all got a chance to learn how to cook some authentic Malay dishes, and I also taught my own improvised version of Avo Chicken (see p. 73 of The *Ham Har Chaan Cookbook*) during the class.

We had dinner the other night at the Majestic Hotel; their restaurant serves Kristang food exclusively. It was a nine-course dinner and the food was awesome! And I noticed the inchimintu karangezu (literally means "baked stuffed cake" with crabmeat and chicken stuffed in crab shell) is prepared very much like my family recipe for Baked Crab Meat in *The Ham Har Chaan Cookbook* (p. 121). And kari debal ("devil's curry") tastes very much like the African Chicken here on p. 213. And I met Chef Melba Nunis, the woman behind this lovely restaurant. We talked about Kristang and Macanese cooking, the use of spices and similarities on ways to prepare pork dishes. I hope I can come back to Malaysia soon and learn some cooking from Chef Melba.

And my room here at Banyan Tree Kuala Lumpur is just stunning. I love the spacious bathroom and a bathtub with a bird's-eye view of the city. I need a hot bath now to relax myself, and I am looking forward to the lovely BBQ wings from the famous W.A.W Restaurant on Jalan Alor tomorrow!

馬葡燉牛舌
Ox Tongue Semur

這道菜是跟隨 Kristang 料理達人 Melba Nunis 學的，她告訴我她的祖父很愛吃燉煮的菜式。Semur 是東南亞的燉肉，將牛舌用濃厚、充滿香草和香料的肉汁燉煮。我將食譜的口味改得比較淡一點，但加入薯仔以豐富質感。

This is a dish I learned to cook from the Kristang chef Melba Nunis. She told me how her grandfather loved to eat the homey and comforting stew and that it is always cooked with love. The *semur* (which means stew) is a Southeast Asian meat stew, with ox tongue browned in a thick gravy sauce with spices and herbs. I have made some changes to the recipe to make it lighter in taste, but richer in texture by adding new potatoes.

材料（4人份量）

橄欖油	2湯匙
洋蔥	150克，切片
肉桂條	1條
黑胡椒粒	1湯匙
牛舌	1公斤，洗淨切方粒
雞湯或牛肉湯	2公升
清水	1公升
豉油	180毫升
老抽	1湯匙
新薯	200克，切方粒
白酒醋	50毫升
乾紅椒碎	2茶匙（可免）
芫茜碎	1茶匙（裝飾用）

Ingredient (Makes 4 servings)

Olive oil	2 tablespoons
White onion, sliced	150g
Cinnamon stick	1
Black peppercorns	1 tablespoon
Ox tongue, cleaned and cut into cubes	1kg
Chicken or beef broth	2 liters
Water	1 liter
Soy sauce	180ml
Dark soy sauce	1 tablespoon
New potatoes, cut into cubes	200g
White wine vinegar	50ml
Dried red chili flakes (optional)	2 teaspoons
Cilantro, chopped	1 teaspoon (for garnish)

做法 How to cook

在湯鍋或燉鍋內以中火加熱橄欖油，放入洋蔥炒至軟身透明。加入肉桂條及黑椒粒炒1分鐘，至有香味釋出。加入牛舌炒5分鐘至微黃。倒入上湯、清水、豉油和老抽，蓋好以中火燜煮30至45分鐘，至牛舌差不多軟腍。放入薯仔和醋再煮15分鐘至薯仔軟腍。喜歡的話可加紅椒碎，灑上芫茜碎裝飾即成。

Heat the olive oil in a saucepan or casserole over medium heat. Add the onion and sauté until soft and translucent. Add the cinnamon stick and peppercorns and stir-fry for 1 minute, until you can smell the fragrance. Add the ox tongue and sauté for 5 minutes, until lightly browned.

Add the broth, water, soy sauce, and dark soy sauce, cover, and simmer over medium heat for 30 to 45 minutes, until the ox tongue is almost tender. Add the potatoes and vinegar and cook for 15 minutes, until the potatoes are soft and tender. You may like to add red chili flakes at the end to finish off. Garnish with chopped cilantro.

馬六甲河畔之家 **Casa del Rio Melaka**

📞 (+606) 289 6888　　　@ www.casadelrio-melaka.com

我下筆之時，馬六甲並沒有太多 5 星級酒店。馬六甲河畔之家就是其中一間，它與有名的雞場街只有 4 分鐘的腳程。

There are not many 5-star hotels in Malacca at the moment I am writing this—Casa del Rio is one of them. It is close to the famous Jonker Street, only a 4-minute walk from the hotel.

吉隆坡悦榕莊 **Banyan Tree Kuala Lumpur**

📞 (+603) 2113 1888　　　@ www.banyantree.com

我很欣賞吉隆坡悦榕莊的客房，甚至普通的豪華客房也相當寬敞。而且浴室設備一流，我可以在這裡浸浴一小時，邊欣賞落地玻璃外的迷人景色。

What I like about Banyan Tree Kuala Lumpur is my guestroom. It's so spacious, even for a deluxe room. And the bathroom is just another WOW—I can spend an hour soaking myself in the tub while enjoying the beautiful view from its floor-to-ceiling window.

The Mansion

📞 (+606) 289 8000 　　@ www.majesticmalacca.com/pages/dining.html

這是馬六甲 Majestic 酒店內的餐廳，提供 Kristang 菜式。

The Mansion is the restaurant serving Kristang cuisine located inside the Majestic Hotel, Malacca.

黃亞華小食店　W.A.W Restaurant S/B

📞 (+603) 2144 2463

位於有名的食街亞羅街。我已光顧了超過十年，你一定要試試這裡的燒雞翼、炒飯和柑桔青檸水，非常美味。

來到亞羅街，不要錯過印度煎餅 Roti canai，是馬來亞西、印尼和新加坡常見的美食，一般配印度雜豆咖喱、魚或咖喱雞食用。但我愛加煉奶及香蕉，無得頂！

Located on the famous food street of Jalan Alor, I have been visiting this restaurant for more than ten years. You must try their roast chicken wings, fried rice, and kumquat and lime juice drink. They are really good.

On Jalon Alor, also look for the Indian *roti canai* store. *Roti canai*, also known as *roti cane* or *roti paratta*, is an Indian-influenced flatbread commonly found in Malaysia, Indonesia, and Singapore. It's usually eaten with dhal, fish, or chicken curry. But I like it served with condensed milk and banana. Thumbs up!

人生如戲，戲如人生。起初還以爲這個烹飪舞台只是曇花一現，轉瞬即逝。但每當我跳高一點，舞台就變得大一點，而這世界也變得大大的。我的烹飪世界確實是無窮無盡，還有很多的東西要嚐、要學習、要體驗。吃過一頓窩心的午餐，執拾行裝，繼續去嚐味。跳高一點、跳遠一點，回味無窮 。

I am so happy to have found a platform that lets me stay longer than I expected. As I jump higher and higher, I can see more and more of the culinary world, a world that is so huge that there is still a lot for me to see, to learn, and to taste. Thank you so much for reading my journal. See you again very soon .

贊助品牌：
Sponsoring Brands :

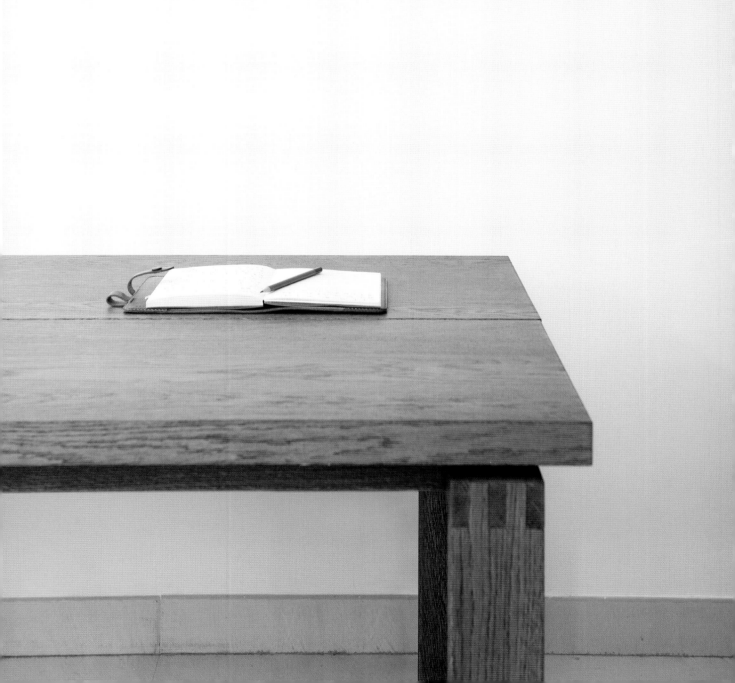

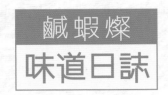

鹹蝦燦
味道日誌

Published in Hong Kong 2019
2019 香港出版

Copyright © John Rocha 2019
Photography copyright © John Rocha 2019

Author: John Rocha
Email: eurasianfoodjourney@gmail.com
Facebook: http://facebook.com/eufoodjourney
Instagram: eurasianfoodjourney
Whatsapp/tel: (852) 5503 8258

Design & Art Direction: Matthew
Photography & Photo Direction: Chau Wai Hung
Videographer & Producer: Nick M
Copy Editor (English): Deri Reed
Managing Editor: Jenny Fung

Published by Ada Wang
出版人：王凱思

WE Press Company Limited
香港人出版有限公司
14/F, Greatmany Centre, 109-115 Queen's Road East, Wan Chai, Hong Kong
香港灣仔皇后大道東109–115號智群商業中心14樓

Website: www.we-press.com   Email: info@we-press.com
Facebook: www.facebook.com/wepresshk
Instagram: we_press   Whatsapp/tel: (852) 6688 1422

ISBN: 978-988-79340-2-8